The Physics of Superconductors

Springer
Berlin
Heidelberg
New York
Barcelona
Budapest
Hong Kong
London
Milan
Paris
Santa Clara
Singapore
Tokyo

P. Müller A.V. Ustinov (Eds.)

V. V. Schmidt

The Physics of Superconductors

Introduction to Fundamentals and Applications

with 114 Figures
and 51 Problems with Solutions

Professor Dr. Paul Müller
Professor Dr. Alexey V. Ustinov
Physikalisches Institut III, Universität Erlangen-Nürnberg
Erwin-Rommel-Str. 1, D-91058 Erlangen, Germany

E-mail
Prof.Müller:phm@physik.uni-erlangen.de
Prof.Ustinov: ustinov@physik.uni-erlangen.de

Translator
Dr. Irina V. Grigorieva
Wildschut, 2
6581 BA Malden
The Netherlands

This book is a completely revised translation of the Russian original edition:
V. V. Schmidt, The Physics of Superconductors, Nauka Publishers, Moskau 1982

Library of Congress Cataloging-in-Publication Data
Schmidt, V. V. (Vadim Vasil 'evich)
[Vvedenie v fiziku sverkhprovodnikov. English]
The physics of superconductors : introduction to fundamentals and applications
V.V. Schmidt ; P. Müller, A.V. Ustinov (eds.) ; [translator, Irina V. Grigorieva]. p. cm.
Includes bibliographical references and index.

1. Superconductors. I. Müller, P. (Paul), 1948- .
II. Ustinov, A. V. (Alexey V.) , 1961- . III. Title.
QC612.S8S5613 1997
537.6 ' 23-dc21

ISBN 978-3-642-08251-1

Preface to the English Edition

The author of this book, Prof. Vadim Vasilievich Schmidt, was known in the West as 'Russian Schmidt'. Being a talented theoretician and at the same time heading an experimental laboratory in Chernogolovka, Vadim Schmidt had a unique talent for explaining complicated physical models and ideas in a simple way. One of his favorite sayings was 'physics is a verbal science'. The book is based on his introductory course on superconductivity. It was published in Russian in 1982 and very soon became undoubtedly the most popular Russian textbook on the physics of superconductors. Without losing its generality and depth, the book presents key aspects of superconductivity in a very clear and logically structured form. As the author said in his preface to the Russian edition, he tried to 'avoid situations when the words "it is easy to show" conceal tedious and bulky computations'. In spite of its relatively small volume, the book gives a broad overview and covers the major topics of superconductivity. Well-selected examples are clearly described and help the reader to understand the ideas presented. The book even allows an inexperienced reader to quickly get a grasp of a topic.

Vadim Schmidt was born in 1927 in Moscow. His father was a well-known politician and the family did not escape the purges of Stalin's time. After spending some of his young years in the Ural region, Vadim Schmidt graduated from the Physics Department of the Moscow State University in 1952. After university, he undertook experimental and theoretical work in the physics of metals. The most fundamental of Schmidt's contributions to the theory of superconductivity was his pioneering work on superconducting fluctuations above T_c which appeared in 1966. At that time it opened a new direction of research. His later works on magnetic flux pinning in superconductors, fluxon dynamics in Josephson junctions, and thermoelectric and other nonequilibrium effects in superconductor–normal metal–superconductor junctions became undoubtedly very important for the field of superconductivity. Schmidt's lectures on superconductors were extremely popular with students and formed the basis of the book.

The Russian edition of this book appeared four years before the discovery of high-temperature superconductivity in 1986. Sadly, the premature death of Vadim Schmidt in 1985 did not allow him to see the beginning of the new era for superconductors. Despite their high critical temperature and several other essential properties, the newly discovered materials remain superconductors in the classic sense which is presented in the book of Schmidt. Published only once in Russian in a small number of copies and at that time not translated into English, this book very soon became a bibliographic rarity. Presently, it can hardly be found even in Russia in any form other than photocopies distributed among the many new students studying superconductivity.

Preparing the book for this first English edition, we decided to keep its main structure and the number of chapters. We thought that, after all, this book should remain a nice introduction to the physics of superconductors in general, but including now high-temperature superconductors. Where appropriate, we added new references and discussions about features specific to high-temperature superconductivity and progress made in the theory, experimental studies, and technology of superconductors during the last decade. In order to be consistent with major textbooks on superconductivity, we have kept the original author's notation in CGS units.

Finally, we acknowledge I. Grigorieva for valuable discussions and suggestions during the preparation of the English edition of the book.

Erlangen, March 1997 *Paul Müller*
Alexey Ustinov

Preface to the Russian Edition

It has been 70 years since the discovery of superconductivity but it is only during the last two decades that superconductors have metamorphosed from exotic objects that physicists use in their experiments into materials of practical importance. New technologies have emerged, where superconducting materials are used for super-high-field magnets or cables capable of loss-free power transmission. Superconductors are becoming increasingly important for the development of high-sensitivity high-precision electronic devices such as bolometers, high-frequency resonators, mixers, and uniquely sensitive devices based on the Josephson effect. Research programs are now under way aimed at developing superconductor-based logic and memory elements for computers.

Due to these developments, a significant number of specialists deal with the phenomenon of superconductivity routinely, on an everyday basis. Appropriate training is now offered at some universities and technical colleges.

At present, several quite general books on superconductivity are available. These are first of all the monographs by A.C. Rose-Innes and E.H. Roderick [1], E.A. Lynton [2], M. Tinkham [3], P.G. de Gennes [4], and D.R. Tilley and J. Tilley [5]. Each of these books is excellent in its own way. However, some of them, such as [3, 4], require a good background in theoretical physics while others do not give a complete up-to-date picture of the physics of superconductors.

This book has evolved from a set of lectures given by the author for a number of years as a course at the Moscow Institute of Steel and Alloys. It was the author's intention that it should mainly give insight into the physics of the phenomenon of superconductivity.

We did our best to avoid situations when the words 'it is easy to show' conceal tedious and bulky computations. Almost all of the results given in the book are derived from 'first principles' so that the reader can follow the computations from the beginning to the end. We always tried to obtain quickly and in a simple way a qualitative result, an estimate of the order of magnitude, instead of describing complex calculations leading to the exact answer.

While writing the book, we assumed that the reader is familiar with the basics of quantum mechanics and the physics of metals.

Chapter 1 is of introductory character. It gives the basic experimental facts, outlines the development of the theory of superconductivity and discusses the thermodynamics of superconductors. Chapter 2 deals with the linear electrodynamics of superconductors based on the London equations. The fundamentals of the Ginzburg–Landau theory of superconductivity make up Chap. 3. Chapter 4 discusses weak superconductivity (Josephson effects). It also describes the basic principles underlying the superconducting quantum interference devices (SQUIDs). Chapter 5 is devoted to type-II superconductors. Chapter 6 presents the basic ideas of the microscopic theory of superconductivity – the Bardeen–Cooper–Schrieffer theory. While going through it, the reader will need only basic knowledge of quantum mechanics. This chapter presents the reader with the ideas of electron–electron interaction via phonons, the ground state of a superconductor, and the energy gap and its temperature dependence. It also shows how the existence of the energy gap results in the possibility of persistent currents. Chapter 7, which concludes the book, presents several studies of nonequilibrium effects in superconductors.

The author is glad to take the opportunity to thank D.A. Kirzhnitz, who read the manuscript prior to publication, for many helpful discussions and remarks.

The author is deeply grateful to K.K. Likharev for discussions on some aspects of the physics of superconductors and for his constructive criticisms of the manuscript.

Moscow–Chernogolovka, 1981/1982 *V. V. Schmidt*

Contents

1. Introduction

1.1 Basic Experimental Facts

1.1.1 The Discovery of Superconductivity

Superconductivity was discovered in 1911 at the Leiden Laboratory. While studying the temperature dependence of the electrical resistivity of mercury, H. Kamerlingh-Onnes discovered that at a temperature T^* in the vicinity of 4 K the resistance of the sample dropped suddenly to zero and remained unmeasurable at all attainable temperatures below T^* [6]. Importantly, as the temperature decreased, the resistance disappeared instantly rather than gradually. It was obvious that the sample had undergone a transformation into a novel, as yet unknown, state characterized by zero electrical resistance. This phenomenon was named 'superconductivity'.

All attempts to find at least traces of resistance in bulk superconductors were to no avail. On the basis of the sensitivity of modern equipment, we can argue that the resistivity of superconductors is zero, at least at the level of $10^{-24}\,\Omega\,\mathrm{cm}$. For comparison, we note that the resistivity of high-purity copper is of the order $10^{-9}\,\Omega\,\mathrm{cm}$ at 4.2 K.

Soon after the discovery of superconductivity in mercury, the same property was found in several other metals: tin, lead, indium, aluminum, niobium, and others. Many alloys and intermetallic compounds also turned out to be superconductors.

The temperature of the transition from the normal to the superconducting state is called the critical temperature T_c. Shortly after the discovery, it was found that superconductivity can be destroyed not only by heating the sample, but also by placing it in a relatively weak magnetic field. This field is called the critical field of bulk material, H_{cm}. In most of the English-language literature, H_{cm} is called the thermodynamic critical field, $H_{c\,th}$.

Table 1.1 gives values of T_c and H_{cm} for a number of superconducting elements. Here $H_{cm}(0)$ is the critical field extrapolated to absolute zero. The temperature dependence of H_{cm} is well described by the empirical formula

$$H_{cm}(T) = H_{cm}(0)\,[1 - (T/T_c)^2]\,. \tag{1.1}$$

Table 1.1. Critical temperatures and critical magnetic fields of superconducting elements [7]

Element	T_c /K	$H_{cm}(0)$ /Oe	Element	T_c /K	$H_{cm}(0)$ /Oe
Al	1.175 ± 0.002	104.9 ± 0.03	Pa	1.4	
Be	0.026		Pb	7.196 ± 0.006	803 ± 1
Cd	0.517 ± 0.002	28 ± 1	Re	1.697 ± 0.006	200 ± 5
Ga	1.083 ± 0.001	59.2 ± 0.3	Ru	0.49 ± 0.015	69 ± 2
Hf	0.128		Sn	3.722 ± 0.001	305 ± 2
Hg (α)	4.154 ± 0.001	411 ± 2	Ta	4.47 ± 0.04	829 ± 6
Hg (β)	3.949	339	Tc	7.8 ± 0.01	1410
In	3.408 ± 0.001	281.5 ± 2	Th	1.38 ± 0.02	160 ± 3
Ir	0.1125 ± 0.001	16 ± 0.05	Ti	0.40 ± 0.04	56
La (α)	4.88 ± 0.02	800 ± 10	Tl	2.38 ± 0.04	178 ± 5
La (β)	6.0 ± 0.1	1096, 1600	V	5.40 ± 0.05	1408
Lu	0.1	< 400	W	0.0154 ± 0.0005	1.15 ± 0.03
Mo	0.915 ± 0.005	96 ± 3	Zn	0.850 ± 0.01	54 ± 0.3
Nb	9.25 ± 0.02	2060 ± 50	Zr	0.61 ± 0.15	47
Os	0.66 ± 0.03	70			

This dependence is shown in Fig. 1.1 which essentially represents the H–T phase diagram of the superconducting state. Within the dashed area, any point in the H–T plane corresponds to the superconducting state.

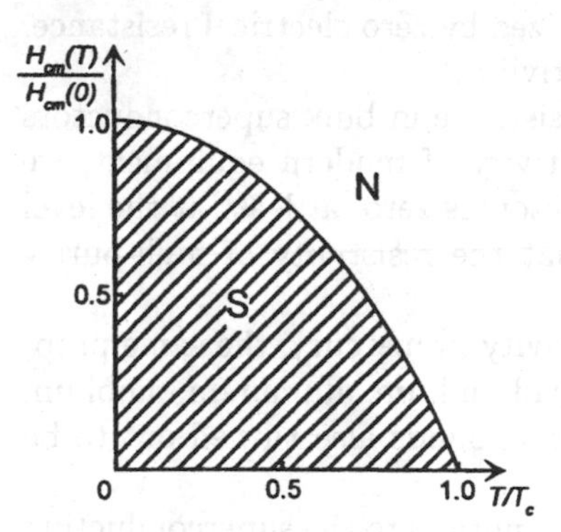

Fig. 1.1. Temperature dependence of the critical field H_{cm}

1.1.2 Magnetic Flux Quantization

An electric current in a superconducting ring can persist for an infinitely long time. Naturally, this does not require a power supply, since the resistance of the ring is zero. Such a persistent current can be produced as follows. Let us place the ring at $T > T_c$ in an external magnetic field so that the magnetic field lines pass through the interior of the ring. Then the ring is cooled down to a temperature below T_c, where the material is superconducting, and the external magnetic field is switched off. At the first moment after switching off the field, the magnetic flux through the ring decreases and, according to Faraday's law of electromagnetic induction, induces a current in the ring which will be persistent from this moment on. This current prevents a further decrease of the magnetic flux through the ring, i.e., now that the external field is zero, the current itself supports the flux through the ring at the initial level. Indeed, if the ring had a finite resistance R, the flux through the ring would decay during the time of the order of L/R, where L is the inductance of the ring. In a superconducting ring, since $R = 0$, it takes the flux infinite time to decay. This means that the magnetic flux becomes 'frozen' while the ring carries a persistent current usually referred to as a superconducting current or a supercurrent.

At first sight it may seem that the 'frozen' magnetic flux can take on an arbitrary value. However, a number of experiments [8, 9] intended to clarify this situation established a very important experimental fact: The magnetic flux through a hollow superconducting cylinder may only assume values that are integral multiples of $\Phi_0 = 2.07 \times 10^{-7}$ G cm^2. The latter is called the magnetic flux quantum and can be written as a combination of fundamental constants: $\Phi_0 = \pi\hbar c/e$, where $\hbar$ is Planck's constant, c is the speed of light and e is the electron charge. In MKSA units $\Phi_0 = h/2e$.

It is surprising that these studies (1961) were not attempted earlier, since by the mid-1930s superconductivity had already been generally perceived as a quantum phenomenon.

1.1.3 Josephson Effects

Another manifestation of the quantum nature of superconductivity was provided by the so-called weak superconductivity, or Josephson, effects [10]. They were predicted in 1962 and soon verified experimentally. The term 'weak superconductivity' refers to a situation in which two superconductors are coupled together by a weak link. The weak link can be provided by a tunnel junction or a short constriction in the cross-section of a thin film. More generally, this can simply be a weak contact between two superconductors over a small area or other arrangements where the superconducting contact between two superconductors is somehow 'weakened'.

There are two Josephson effects to distinguish: stationary (the dc Josephson effect) and nonstationary (the ac Josephson effect).

Consider first the dc effect. Let us apply a current through a weak link (or, in other words, through a Josephson junction). Then, if the current is sufficiently small, it passes through the weak link without resistance, even if the material of the weak link itself is not superconducting (for example, if it is an insulator in a tunnel junction). Here we directly come across the most important property of a superconductor: the coherent behavior of superconducting electrons. Through the weak link, the electrons of the two superconductors merge into a single quantum body. The same can be said in a different way. Having penetrated via the weak link into the second superconductor, the wavefunction of electrons from the first superconductor interferes with the 'local' electron wavefunction. As a result, all superconducting electrons on both sides of the weak link are described by the same wavefunction. The presence of the weak link should not change significantly the wavefunctions on the two sides, compared to what they had been before the link was established.

The ac Josephson effect is even more remarkable. Let us increase the dc current through the weak link until a finite voltage appears across the junction. Then, in addition to a dc component V, the voltage will also have an ac component of angular frequency ω, so that

$$\hbar\omega = 2eV . \tag{1.2}$$

A fundamental experiment intended to record this so-called Josephson radiation (i.e., the electromagnetic radiation emitted by a Josephson junction) was successfully carried out by I.K. Yanson, V.M. Svistunov, and I.M. Dmitrenko [11].

1.1.4 The Meissner–Ochsenfeld Effect

For as long as 22 years after the discovery of superconductivity, scientists believed that a superconductor was simply an ideal conductor, that is, a piece of metal with zero resistance.

Let us work out how such an ideal conductor should behave in an external magnetic field that is weak enough so as not to destroy the specimen's ideal conductivity.

Suppose that initially the ideal conductor is cooled down below the critical temperature in zero external magnetic field. After that an external field is applied. From general considerations, it is easy to show that the field does not penetrate the interior of our sample, as shown schematically in Fig. 1.2. Indeed, immediately after the field penetrates the surface layer of the ideal conductor, an induced current is set up which, according to Lenz's law, generates a magnetic field in the direction opposite to that of the external field. Therefore, the total field in the interior of the specimen is zero.

Let us prove this with the help of Maxwell's equations. As the induction $\boldsymbol{B}$ changes, an electric field $\boldsymbol{E}$ must be induced in the specimen with

$$\operatorname{curl} \boldsymbol{E} = -\frac{1}{c}\frac{\partial \boldsymbol{B}}{\partial t} ,$$

where c is the speed of light in vacuum. In the ideal conductor $\boldsymbol{E} = 0$, since $\boldsymbol{E} = \boldsymbol{j}\rho$, where ρ is the resistivity (which in our case is zero) and $\boldsymbol{j}$ is the density of the induced current. It follows that $\boldsymbol{B} =$ const and, taking into account that $\boldsymbol{B} = 0$ before applying the external field, we arrive at $\boldsymbol{B} = 0$ also after the field is applied. This phenomenon can also be interpreted in a different way: since $\rho = 0$, the time for the magnetic field penetration into an ideal conductor is infinitely long.

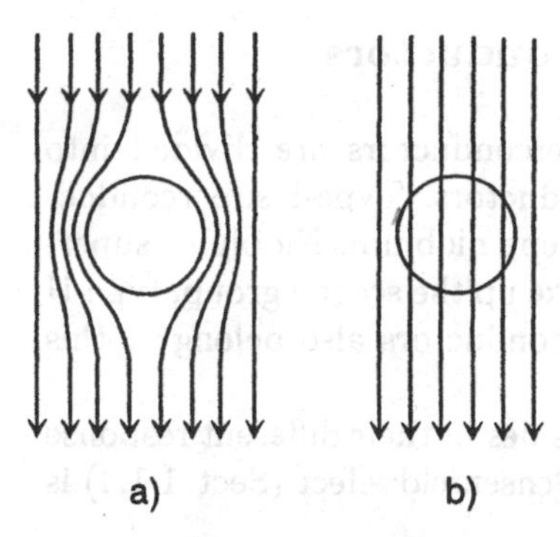

Fig. 1.2. For an ideal conductor, its magnetic state at $T < T_c$, $H > 0$ depends on its history: (a) magnetic field applied to an ideal conductor at $T < T_c$; (b) field applied at $T > T_c$

Thus, we have established that $\boldsymbol{B} = 0$ at any point of an ideal conductor placed in an external magnetic field. However, the same situation (an ideal conductor at $T < T_c$ in an external magnetic field) can be reached through a different sequence of events, that is, by first applying the external field to a 'warm' specimen and then cooling it down to $T < T_c$.

In that case, for the ideal conductor, electrodynamics predicts an entirely different result. At $T > T_c$, the resistivity of the specimen is finite and, therefore, the magnetic field penetrates into it. After cooling the specimen down through the superconducting transition, the field remains in it, as illustrated in Fig. 1.2 (b).

Note that in the above reasoning we always referred to the specimen characterized by $\rho = 0$ as an ideal conductor and not a superconductor.

It was generally believed before 1933 that the superconductor was indeed simply an ideal conductor. But the experiment by W. Meissner and R. Ochsenfeld [12] revealed that this was not true! They found that at $T < T_c$ the field inside a superconducting specimen was always zero ($\boldsymbol{B} = 0$) in the presence of an external field, independent of which procedure had been chosen to cool the superconductor through T_c.

This discovery was immensely important. Indeed, if $\boldsymbol{B} = 0$ independent of the specimen's history, the zero induction can be treated as an intrinsic property of the superconducting state at $H < H_{cm}$. Furthermore, it implies that we can treat a transition to the superconducting state as a phase transition and, consequently, apply all the might of the thermodynamic approach to examine the superconducting phase.

Thus, the superconducting state obeys the equations

$$\rho = 0\,, \tag{1.3}$$

$$\boldsymbol{B} = 0\,. \tag{1.4}$$

1.2 Magnetic Properties of Superconductors

According to their magnetic properties, superconductors are divided into type-I superconductors and type-II superconductors. Type-I superconductors include all superconducting elements except niobium. Niobium, superconducting alloys and chemical compounds make up the second group, type-II superconductors. The so-called high-T_c superconductors also belong to this group.

The chief difference between the two groups lies in their different response to an external magnetic field. The Meissner–Ochsenfeld effect (Sect. 1.1.4) is observed in type-I superconductors only.

1.2.1 Magnetic Properties of Type-I Superconductors

Let us consider the magnetization curve of a superconductor. Suppose that the specimen is a long cylinder in a longitudinal external magnetic field $\boldsymbol{H}_0$. As the field $\boldsymbol{H}_0$ increases, the induction inside the specimen does not change at first; it remains $\boldsymbol{B} = 0$. As soon as H_0 reaches the value of H_{cm}, the superconductivity is destroyed, the field penetrates into the superconductor, and $\boldsymbol{B} = \boldsymbol{H}_0$. Therefore, the magnetization curve $B = B(H_0)$ appears as shown in Fig. 1.3 (a). The magnetic induction $\boldsymbol{B}$ and the magnetic field $\boldsymbol{H}_0$ are related to each other by the well-known expression

$$\boldsymbol{B} = \boldsymbol{H}_0 + 4\pi \boldsymbol{M}, \tag{1.5}$$

where $\boldsymbol{M}$ is the magnetic moment per unit volume. The magnetization curve is often plotted as $-4\pi M$ versus H_0, as illustrated in Fig. 1.3 (b). We shall now derive the basic magnetic properties of type-I superconductors proceeding from equations (1.3) and (1.4) [13].

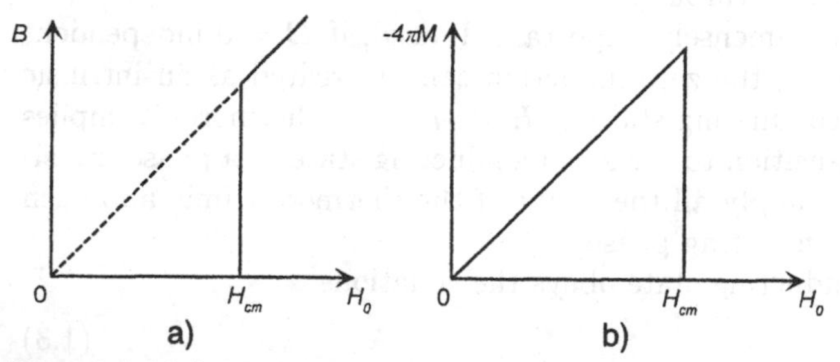

Fig. 1.3. **(a)** Magnetization curve of a superconductor; **(b)** magnetic moment per unit volume, M, versus H_0

(1) *Magnetic field lines outside a superconductor are always tangential to its surface.* Indeed, we know from electrodynamics that magnetic field lines, i.e., lines of the magnetic induction $\boldsymbol{B}$, are continuous and closed. This can

be written as the equation

$$\mathrm{div}\boldsymbol{B} = 0\,,$$

from which it follows that the components of $\boldsymbol{B}$ normal to the surface must be equal on both sides of the surface, that is, inside and outside a piece of material. However, the field in the interior of a superconductor is $\boldsymbol{B}^{(\mathrm{i})} = 0$ and, consequently, $\boldsymbol{B}_{\mathrm{n}}^{(\mathrm{i})} = 0$. It follows that the normal component $\boldsymbol{B}_{\mathrm{n}}^{(\mathrm{e})}$ at the outside of the superconductor's surface is zero, too: $\boldsymbol{B}_{\mathrm{n}}^{(\mathrm{e})} = 0$. And the meaning of the last equation is exactly that the magnetic field lines are tangential to the surface of the superconductor.

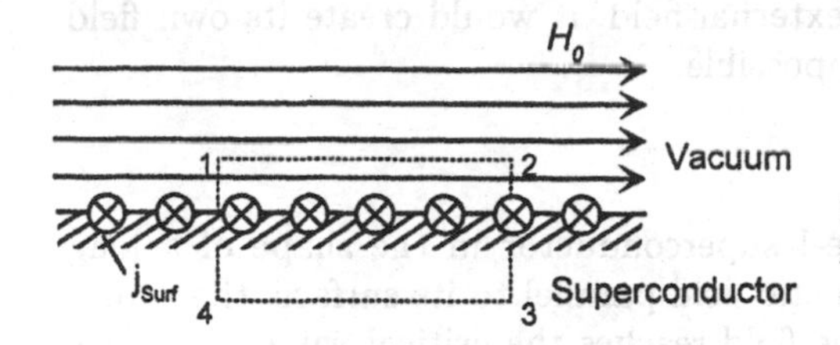

Fig. 1.4. A superconductor placed in an external magnetic field carries a surface current j_{surf}

(2) One of the consequences of the previous property is that *a superconductor in an external magnetic field always carries an electric current near its surface*. This statement is illustrated in Fig. 1.4. Let $\boldsymbol{H}_0$ be the field at a certain point of the surface of a superconductor. Then it follows from Maxwell's equation $\mathrm{curl}\,\boldsymbol{B} = (4\pi/c)\,\boldsymbol{j}$ and the requirement $\boldsymbol{B} = 0$ that the volume current in the interior of the superconductor is zero ($\boldsymbol{j} = 0$) and only a surface current is possible. To elucidate this statement, let us consider the contour *1–2–3–4–1* in Fig. 1.4. First we find the circulation of $\boldsymbol{B}$ about this contour: $\oint \boldsymbol{B}\,\mathrm{d}\boldsymbol{l}$. Along the section *1–2*, which is parallel to the surface, we have $\int_1^2 \boldsymbol{B}\,\mathrm{d}\boldsymbol{l} = H_0 l_{12}$, where l_{12} is the length of the section *1–2*. The contributions from the sections *2–3* and *1–4* are both zero because, from symmetry arguments, the vector $\boldsymbol{B}$ in these two sections is orthogonal to the integration path. The contribution from the section *3–4* is zero, too, because $\boldsymbol{B} = 0$ in the interior of the superconductor. The result is $\oint \boldsymbol{B}\,\mathrm{d}\boldsymbol{l} = H_0 l_{12}$. On the other hand, we find from Maxwell's equation

$$\oint \boldsymbol{B}\,\mathrm{d}\boldsymbol{l} = \frac{4\pi}{c} I\,,$$

where I is the total current through the surface bounded by the contour *1–2–3–4–1*. It follows that there must be a surface current at the surface of the superconductor, flowing 'into the page' (as shown in Fig. 1.4) and of linear density j_{surf} defined by

$$H_0 l_{12} = \frac{4\pi}{c}\, j_{\mathrm{surf}}\, l_{12}\,.$$

The relation between the surface current and the magnetic field at the surface of the superconductor can be written

$$\boldsymbol{j}_{\text{surf}} = \frac{c}{4\pi}\left[\boldsymbol{n} \times \boldsymbol{H}_0\right], \qquad (1.6)$$

where $\boldsymbol{n}$ is the unit vector along the normal to the surface.

Thus, the surface current $\boldsymbol{j}_{\text{surf}}$ is completely defined by the magnetic field $\boldsymbol{H}_0$ at the surface of a superconductor. In other words, the surface current automatically assumes a value such that the magnetic field generated by it inside a superconductor is exactly equal in value and opposite in direction to the external field. This assures zero total field in the interior: $\boldsymbol{B} = 0$.

(3) Let us point out one more, almost obvious, property: *In a simply connected superconductor*[1] *surface currents can exist only when the superconductor is placed in an external magnetic field.* Indeed, if the surface current remained after switching off the external field, it would create its own field in the superconductor which is impossible.

1.2.2 The Intermediate State

We already know that for a type-I superconductor in the shape of a long cylinder placed in a uniform magnetic field parallel to its surface, the superconductivity is destroyed when the field reaches the critical value H_{cm}. It is much harder, however, to destroy the superconductivity of the same cylinder if it is placed in a transverse magnetic field. The same is true for an ellipsoid, a sphere, or other bodies of more complicated geometries.

Consider, for example, the behavior of a superconducting sphere placed in an external magnetic field (Fig. 1.5). Since the magnetic field lines are always tangential to the surface of a superconductor, for our sphere it is obvious that the field lines have a higher density at the 'equator' thereby producing a local increase of the magnetic field. At the same time, the field at the 'poles' is zero. Far away from the sphere, where any perturbations caused by it average out, the uniform external field H_0 is lower than that at the 'equator'.

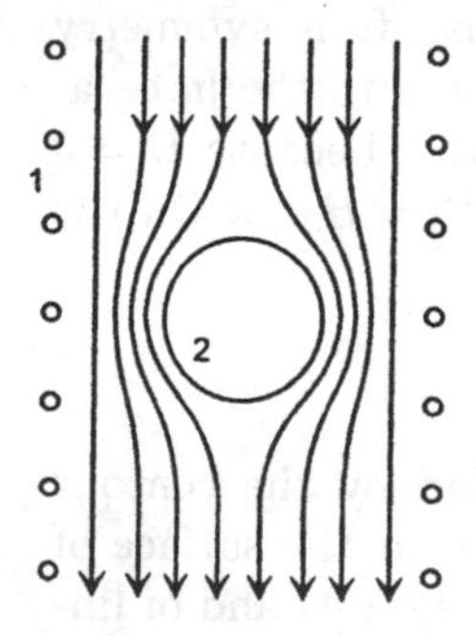

Fig. 1.5. Superconducting sphere in the homogeneous field of a solenoid; *1* – the winding of the solenoid, *2* – the superconducting sphere

A natural question now arises: what happens when the equatorial field reaches the critical value, H_{cm}? Clearly, H_0 at that moment is below H_{cm}

[1] A simply connected body refers to a body inside which an arbitrary closed path can be reduced to a point without having to cross the boundaries of the body.

and, therefore, it is not permitted for the whole sphere to revert to the normal state. On the other hand, it is not permitted for the whole sphere to be superconducting either, because the field at the 'equator' has already reached the critical value.

This contradiction is reconciled by a coexistence of alternating superconducting and normal regions within the sphere which is called the intermediate state. The interfaces between these regions are always parallel to the external field, while in the cross-section perpendicular to the field they may assume various intricate configurations.

Assume that before a superconducting body goes into the intermediate state, the maximum field at its surface (in the case of a sphere, at the 'equator') is H_m and the external field far away from the body is H_0. It is clear then that, on the one hand, $H_m > H_0$ and, on the other hand, H_m is proportional to H_0, with the proportionality coefficient dependent on the exact shape of the body. This can be written in the form

$$H_m = \frac{H_0}{1-n}. \tag{1.7}$$

The values of n for various geometries[2] are given in Table 1.2.

Table 1.2. The demagnetizing factor n for various geometries

Sample geometry	n
Cylinder in parallel field	0
Cylinder in transverse field	1/2
Sphere	1/3
Thin plate in perpendicular field	1

With the help of Table 1.2 we can readily calculate the field H_0 corresponding to the transition into the intermediate state for a body of a certain shape. The transition occurs when the field H_m reaches the value of H_{cm}. Therefore, for a sphere it takes place when, from (1.7), the external field H_0 reaches the value of

$$H_0 = H_{cm}\,(1 - 1/3) = 2/3\,H_{cm}\,.$$

A thin slab in a perpendicular magnetic field shows the intermediate state starting from an infinitesimal magnetic field H_0. This follows both from (1.7) ($n = 1$) and from general physical arguments, as we shall now demonstrate. Consider a thin disk in a perpendicular magnetic field. Since the magnetic field lines must go around the disk, their density increases substantially at the disk edges. The larger the radius of the disk, the larger the increase of

[2] The number n is often called the demagnetizing factor.

the field line density, i.e., of the field H_m, will be. For a disk of infinite radius, the transition to the intermediate state occurs in an infinitesimal field H_0.

A thin slab in the intermediate state is shown in Fig. 1.6.

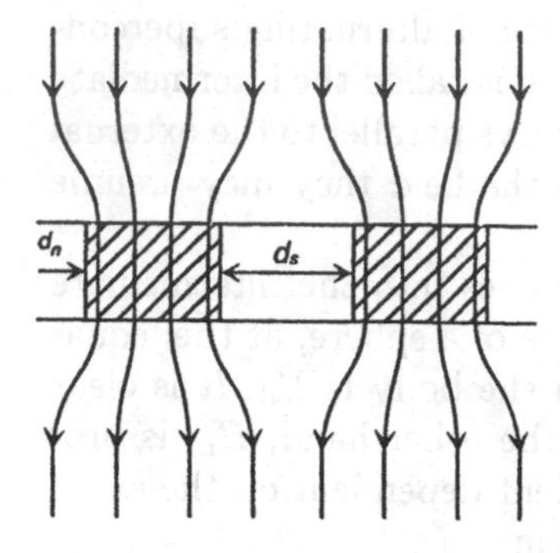

Fig. 1.6. Intermediate state of a thin superconducting plate in a perpendicular field; d_s is the width of a superconducting region, d_n is the width of a normal region

Let us now consider the conditions for thermodynamic equilibrium in the intermediate state.

We shall show that the field in the normal regions always equals the critical field H_{cm}, and these regions automatically adjust their size d_n so as to provide the correct value of the field (see Fig. 1.6). Assume that the field in one of the normal regions exceeds H_{cm}. Then it must destroy the superconductivity of the adjacent superconducting regions. Conversely, if the field in a normal region is less than H_{cm}, this region must be superconducting. Therefore, a stable coexistence of the normal and superconducting regions is only possible if the field in the normal regions equals H_{cm}.

In the cross-section perpendicular to the magnetic field, the domain patterns of the intermediate state can be very irregular. Classic experiments intended to clarify their configurations were performed by A.I. Shalnikov and A.G. Meshkovskii [14, 15]. They used a thin bismuth wire to study the magnetic field distribution in the equatorial cross-section of a tin sphere. In order to do this, they cut the sphere along the equatorial plane and then placed the two half-spheres slightly apart, with a very narrow gap between them, so that the external field was perpendicular to the surfaces of the cut. Then a bismuth wire was inserted into the gap. At low temperatures, the resistivity of the wire depends strongly on the external magnetic field and can be used as a probe of the field. By moving this probe inside the gap it was possible to examine the magnetic flux pattern. The result is shown in Fig. 1.7. The intermediate state was also studied by other methods, for example, by sputtering tiny ferromagnetic particles onto the superconducting surface and thereby producing a replica of the magnetic flux pattern (Fig. 1.8).

The intermediate state assumes an interesting and unusual pattern in a cylindrical wire carrying an electric current. As soon as the current reaches a value such that the magnetic field generated by it at the surface of the wire equals H_{cm}, the wire starts to show the intermediate state. A sketch of the distribution of normal (blank) and superconducting (shaded) regions in such a wire is shown in Fig. 1.9, for a current exceeding the critical value.

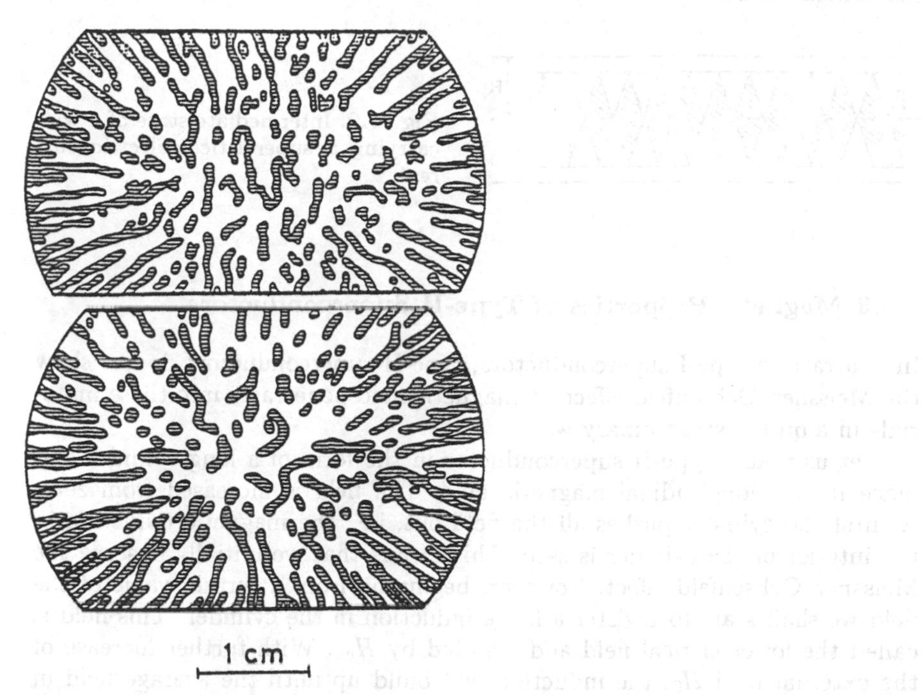

Fig. 1.7. Superconducting and normal regions in a tin sphere [15]. Shaded regions are superconducting

Fig. 1.8. Intermediate state of a single-crystalline tin foil in perpendicular magnetic field [16]; the foil thickness is 2.9×10^{-5} cm

As one can see, a normal layer of thickness $(R - a)$ forms at the surface of the wire, and its thickness grows in proportion to the excess current, above the critical value.

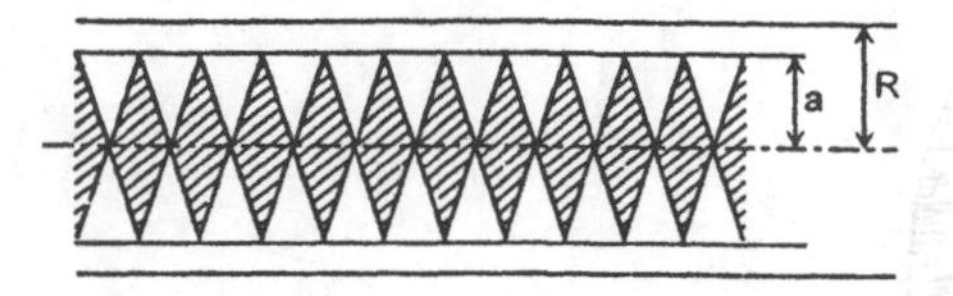

Fig. 1.9. Intermediate state of a wire carrying a supercritical electric current

1.2.3 Magnetic Properties of Type-II Superconductors

In contrast to type-I superconductors, type-II superconductors do not show the Meissner–Ochsenfeld effect. A magnetic field penetrates into these materials in a quite extraordinary way.

Let us take a type-II superconductor in the form of a long cylinder and place it in a longitudinal magnetic field. The field is increased from zero. At first the cylinder pushes all the field out, i.e., the magnetic induction in the interior of the cylinder is zero. This means that we initially observe the Meissner–Ochsenfeld effect. However, beginning from a certain value of the field we shall start to register a finite induction in the cylinder. This field is called the lower critical field and denoted by H_{c1}. With further increase of the external field H_0, the induction will build up until the average field in the cylinder becomes equal to the external field H_0 and the cylinder goes to the normal state. This will happen at the so-called upper critical field H_{c2}. However, in a thin surface layer the superconductivity will remain even at $H_0 > H_{c2}$, until $H_0 = 1.69\,H_{c2}$. This field is strong enough to destroy the superconductivity in the surface layer as well. It is called the third critical field and denoted by H_{c3}.

The physics of type-II superconductors will be discussed in more detail in Chap. 5.

1.3 Thermodynamics of Superconductors

1.3.1 Critical Field of Bulk Material. Thermodynamic Critical Field

Consider a long cylinder of a type-I superconductor in a uniform longitudinal magnetic field H_0. We shall now work out the value of the field which destroys its superconductivity, that is, we shall find the value of H_{cm}.

At $H_0 < H_{cm}$, $\boldsymbol{B}$ is zero due to the Meissner–Ochsenfeld effect. The magnetic moment $\boldsymbol{M}$ per unit volume of the cylinder is then

$$\boldsymbol{M} = -\boldsymbol{H}_0/4\pi \; . \tag{1.8}$$

When a value $\mathrm{d}\boldsymbol{H}_0$ is added to the magnetic field $\boldsymbol{H}_0$, an external source of magnetic field does a work on the superconductor, per unit volume, of

$$-\boldsymbol{M}\,\mathrm{d}\boldsymbol{H}_0 = \boldsymbol{H}_0\,\mathrm{d}\boldsymbol{H}_0/4\pi \; . \tag{1.9}$$

Consequently, when the field changes from 0 to H_0, the work done by the field source is

$$-\int_0^{H_0} \boldsymbol{M}\,\mathrm{d}\boldsymbol{H}_0 = {H_0}^2/8\pi \; . \tag{1.10}$$

This work is stored in the free energy of the superconductor placed in the field H_0. Thus, if the free energy density of a superconductor in zero magnetic field is F_{s0}, that of the superconductor in a finite magnetic field is

$$F_{sH} = F_{s0} + {H_0}^2/8\pi \; . \tag{1.11}$$

A transition to the normal state occurs when the free energy F_{sH} reaches the level of the free energy of the normal metal, i.e., $F_{sH} = F_{\mathrm{n}}$ at $H_0 = H_{\mathrm{cm}}$. This means that

$$F_{\mathrm{n}} - F_{s0} = H_{\mathrm{cm}}^2/8\pi \; . \tag{1.12}$$

It follows from (1.12) that the critical field of a bulk superconductor is a measure of the extent to which the superconducting state is preferable to the normal one from the standpoint of free energy. In other words, it measures the difference in free energy between the normal and superconducting states. The field H_{cm} is often called the thermodynamic critical field.

1.3.2 Entropy of a Superconductor

By the first law of thermodynamics we have

$$\delta Q = \delta R + \delta U \; , \tag{1.13}$$

where δQ is the element of the thermal energy density for the body under consideration, δR is the work done by the body on external bodies, per unit volume, and δU is the element of its internal energy. By definition, the free energy density F is

$$F = U - TS \; , \tag{1.14}$$

where T is the temperature of the body and S is the entropy per unit volume. Then

$$\delta F = \delta U - T\delta S - S\delta T \; .$$

Since for a reversible process $\delta Q = T\delta S$, we have

$$\delta U = T\delta S - \delta R \; , \tag{1.15}$$

$$\delta F = -\delta R - S\delta T \,. \tag{1.16}$$

It follows from (1.16) that

$$S = -(\partial F/\partial T)_R \,. \tag{1.17}$$

Let us use (1.17) to calculate the difference in entropy between the normal and superconducting states. Substituting (1.12) into (1.17) we get

$$S_s - S_n = \frac{H_{cm}}{4\pi}\left(\frac{\partial H_{cm}}{\partial T}\right)_R \,. \tag{1.18}$$

From (1.18), important physical results can be derived.

(1) According to Nernst's theorem, the entropy of a body at $T = 0$ is zero. Therefore $(\partial H_{cm}/\partial T)_{T=0} = 0$, i.e., at $T = 0$ the curve $H_{cm}(T)$ has zero derivative.
(2) It is known from experiment that, as the temperature T increases, the dependence $H_{cm}(T)$ has the form of a monotonically descending curve (see Fig. 1.1). That is, for the entire temperature interval from 0 to T_c, we have $\partial H_{cm}/\partial T < 0$ and $S_s < S_n$.
(3) Since $H_{cm} = 0$ at $T = T_c$, we get $S_s = S_n$ at $T = T_c$. A sketch of the $S_s - S_n$ dependence on temperature is shown in Fig. 1.10.

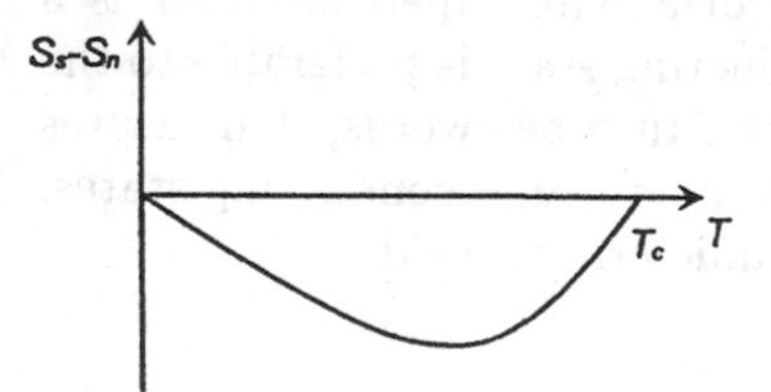

Fig. 1.10. Temperature dependence of the entropy difference $S_s - S_n$

We now arrive at some very important conclusions.

(1) The superconducting state is a more ordered state than the normal one because it is characterised by lower entropy.
(2) Since $S_s = S_n$ at $T = T_c$, the transition at $T = T_c$ does not involve latent heat. Therefore, the transition at $T = T_c$ is a second-order phase transition.
(3) At $T < T_c$, a transition from the superconducting to the normal state occurs when a sufficiently strong magnetic field is applied. Since, as we know, $S_s < S_n$, such a transition is accompanied by absorption of latent heat. For the transition from the superconducting to the normal state, it is the other way round, i.e., the latent heat is given up. Therefore, in the presence of a magnetic field, all the transitions at $T < T_c$ are first-order phase transitions.

It is amazing how a couple of formulas from thermodynamics and a single experimental fact – the dependence of H_{cm} on T (Fig. 1.1) – have led to such fundamental conclusions! Fundamental indeed, because from our new knowledge that the superconducting state is more ordered (i.e., possesses a lower entropy) than the normal one, we come straight to the understanding that superconductivity is based on the coherent behavior of electrons.

1.3.3 Heat Capacity

The specific heat of matter can be defined as $C = T(\partial S/\partial T)$. Therefore, we can write the difference in specific heat between the superconducting and normal states as

$$C_s - C_n = \frac{T}{4\pi}\left[\left(\frac{\partial H_{cm}}{\partial T}\right)^2 + H_{cm}\frac{\partial^2 H_{cm}}{\partial T^2}\right] . \qquad (1.19)$$

In order to obtain (1.19), we have taken the derivative of (1.18). Now, since $H_{cm} = 0$ at $T = T_c$, we have

$$C_s - C_n = \frac{T_c}{4\pi}\left(\frac{\partial H_{cm}}{\partial T}\right)^2_{T_c} . \qquad (1.20)$$

Expression (1.20), known as the Rutgers formula, implies that there must be a discontinuous jump at $T = T_c$ in the specific heat as a function of temperature. The Rutgers formula defines the height of the jump. The temperature dependence of the specific heat is plotted in Fig. 1.11. At $T > T_c$, the specific heat decreases linearly with decreasing temperature, as is usually the case for normal metals (electronic specific heat).

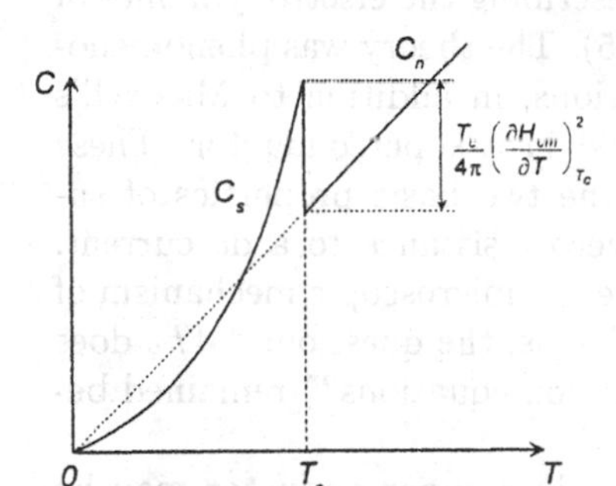

Fig. 1.11. Temperature dependence of the specific heat of a superconductor

For the moment we shall consider the form of the $C(T)$ dependence in Fig. 1.11 simply as an experimental fact, i.e., not supported by theoretical analysis. However, the fact that there exists a point at $T < T_c$ where the two curves – $C_s(T)$ and the extrapolation of $C_n(T)$ – intersect, follows from the above argumentation. Indeed, at this point $C_s = C_n$, that is, $\partial(S_s - S_n)/\partial T = 0$. And that such a point must exist follows from Fig. 1.10.

1.3.4 Free Energy

Let us consider in more detail the basic properties of those thermodynamic potentials that we shall often deal with when studying superconductivity. These potentials are equally suitable for analysis of any sort of matter, superconductors in particular. Definitions and detailed arguments can be found in [13].

Consider a body in an external magnetic field H_0 such that its temperature and the magnetic induction in its interior are constant. The body attains thermodynamic equilibrium when the free energy $\mathcal{F} = \int F \, \mathrm{d}V$ is a minimum, where the free energy density F is defined by (1.14). The quantity $\mathcal{F}$ is sometimes referred to as the Helmholtz free energy.

For many calculations, making use of this potential is awkward because normally the quantities presumed to be constant for a body in an external magnetic field are the temperature and the external magnetic field H_0. In the latter case, thermodynamic equilibrium is attained when another thermodynamic potential is a minimum, that is, the Gibbs free energy $\mathcal{G}$ defined as

$$\mathcal{G} = \int G \, \mathrm{d}V \,, \tag{1.21}$$

$$G = F - \boldsymbol{B} \cdot \boldsymbol{H}_0 / 4\pi \,. \tag{1.22}$$

1.4 Essay on the Development of the Theory of Superconductivity

The first theory which proved successful in describing the electrodynamics of superconductors was the London theory (1935). The theory was phenomenological, that is, it had introduced two equations, in addition to Maxwell's equations, governing the electromagnetic field in a superconductor. These equations provided a correct description of the two basic properties of superconductors: absolute diamagnetism and zero resistance to a dc current. The London theory did not attempt to resolve the microscopic mechanism of superconductivity on the level of electrons, that is, the question: "*Why* does a superconductor behave according to the London equations?" remained beyond its scope.

According to the London theory, electrons in a superconductor may be considered as a mixture of two groups: superconducting electrons and normal electrons. The number density of the superconducting electrons, n_s, decreases with increasing temperature and eventually becomes zero at $T = T_c$. At $T = 0$ it is the other way round, that is, n_s is equal to the total density of conduction electrons. These are the postulates of the so-called two-fluid model of a superconductor. A flow of superconducting electrons meets no resistance. Such a current, obviously, cannot generate a constant electric field in a superconductor because, if it did, it would cause the superconducting electrons to accelerate infinitely. Therefore, under stationary conditions, that is, without an electric field, the normal electrons are at rest. In contrast, in the presence of an ac electric field, both the normal and the superconducting components of the current are finite and the normal current obeys Ohm's law.

In the framework of this concept, a real superconductor can be modeled by an equivalent circuit consisting of a normal resistor and an ideal conductor connected in series. The ideal conductor in the circuit must have a finite inductance to mimic the inertia of the superconducting electrons.

The London equations provided a description for the behavior of the superconducting component of the electronic liquid in both dc and ac electromagnetic fields. They also helped to understand a number of aspects of the superconductors' behavior in general. However, by the end of the 1940s, it was clear that to one question at least the London theory gave a wrong answer. For an interface between adjacent normal and superconducting regions, the theory predicted a negative surface energy $\sigma_{ns} < 0$. This implied that a superconductor in an external magnetic field could decrease its total energy by turning into a mixture of alternating normal and superconducting regions. In order to make the total area of the interface within the superconductor as large as possible, the size of the regions must be as small as possible. This was supposed to be the case even for a long cylinder in a longitudinal magnetic field, in contradiction to experimental evidence existing at that time. Experiment showed that such a separation of the normal and superconducting regions occurred only for specimens having a nonzero demagnetizing factor (the intermediate state, see Sect. 1.2.2). In addition, the layers were rather thick (~ 1 mm as in Fig. 1.7), which could only be the case if $\sigma_{ns} > 0$, again in contradiction to the predictions of the London theory.

The above contradiction was reconciled by a theory proposed by V.L. Ginzburg and L.D. Landau (the so-called GL theory) which was also phenomenological but took account of quantum effects. Why it is so important to include quantum effects will become clear in a moment. Assume that there exists a wavefunction Ψ describing the superconducting electrons quantum-mechanically. Then the squared amplitude of this function (which is proportional to n_s) must be zero in a normal region, increase smoothly through the normal–superconducting (NS) interface and finally reach a certain equilibrium value in a superconducting region. Therefore, a gradient of Ψ must appear at the interface. At the same time, as is well known from quantum mechanics, $|\nabla\Psi|^2$ is proportional to the density of the kinetic energy. Thus, by taking into account quantum effects, we also take into account an additional positive energy stored at the NS interface, which creates the opportunity to obtain $\sigma_{ns} > 0$.

The GL theory will be discussed in detail in Chap. 3. Here we shall only outline its most important ideas which have determined the tremendous importance of the GL theory for the entire field of superconductivity.

The GL theory introduced quantum mechanics into the description of superconductors. It assigned to the entire number of superconducting electrons a wavefunction depending on a single spatial coordinate [recall that, generally speaking, a wavefunction of n electrons in a metal is a function of n coordinates, $\Psi(\boldsymbol{r}_1, \boldsymbol{r}_2, \ldots, \boldsymbol{r}_n)$]. By doing so, the theory established the coherent behavior of all superconducting electrons. Indeed, in quantum mechanics, a single electron in the superconducting state is described by a function $\Psi(\boldsymbol{r})$. If we now have n_s absolutely identical electrons (where n_s, the superconducting electron number density, is a macroscopically large number) and all these electrons behave coherently, it is clear that the same wavefunction of a single parameter is sufficient to describe all of them. This idea was a breakthrough that enabled the prediction of many beautiful quantum, and at the same time macroscopic, effects in superconductivity.

The GL theory was built on the basis of the theory of second-order phase transitions (the Landau theory) and, therefore, it is valid only in the vicinity of the critical temperature, that is, within the temperature range $T_c - T \ll T_c$. The scope of its validity will be discussed in more detail in Sect. 6.6.

Having applied the GL theory to superconducting alloys, A.A. Abrikosov developed a theory of the so-called type-II superconductors (1957). It turned out that superconductors do not necessarily have to have $\sigma_{ns} > 0$. Those that do have $\sigma_{ns} > 0$ are type-I superconductors. But the majority of superconducting alloys and chemical compounds show $\sigma_{ns} < 0$; they are type-II superconductors. For type-II superconductors, there is no Meissner effect and the magnetic field does penetrate inside the material, but penetrates in a very unusual way, that is, in the form of quantized vortex lines (quantum effect on the macroscopic scale!). Superconductivity in these materials can survive up to very high magnetic fields.

Still, neither the London, nor the GL theory could answer the question: "What *are* those 'superconducting electrons' whose behavior they were intended to describe?" It was 46 years since the discovery of superconductivity but at the microscopic level a superconductor remained a mystery.

This question was finally resolved in 1957 by the work of J. Bardeen, L. Cooper and J. Schrieffer (the BCS theory). An important contribution was also made by N.N. Bogolyubov (1958) who developed a mathematical method now widely used in studies of superconductivity.

The decisive step in understanding the microscopic mechanism of superconductivity is due to L. Cooper (1956). The essence of his work can be outlined as follows. Consider a normal metal in the ground state: in $\boldsymbol{k}$ space, all states for non-interacting electrons inside the Fermi sphere are occupied, while all those outside it are empty. Then an extra pair of electrons is brought in and placed in the states $(\boldsymbol{k}\uparrow)$ and $(-\boldsymbol{k}\downarrow)$, in the vicinity of the Fermi surface (the arrows indicate the directions of electron spins). It turned out that if, for whatever reason, the two electrons become attracted to each other, they form a bound state regardless of how weak the attraction is. In real space, these electrons form a bound pair – a Cooper pair.

What the BCS theory has demonstrated is that taking into account the interaction between electrons and phonons can, under certain circumstances, lead to electron–electron attraction. As a result, a part of the electrons form Cooper pairs. The total spin of a Cooper pair is zero, which means that it represents a Bose particle (that is, obeys Bose–Einstein statistics). Such particles possess a remarkable property: if the temperature of a system falls below a certain temperature T_c, they can all gather at the lowest energy level (in the ground state). Furthermore, the larger the number of the particles that have accumulated there, the more difficult it is for one of them to leave this state. This process is called Bose condensation. All the particles in the condensate have the same wavefunction, depending on a single spatial coordinate. One can easily understand that the flow of such a condensate must be superfluid, that is, dissipation-free. Indeed, it is not easy at all for one of the particles to be scattered by, say, an impurity atom or by any other defect of the crystal lattice. In order to become scattered, the particle would first have to overcome the 'resistance' of the rest of the condensate.

Thus one can briefly describe the phenomenon of superconductivity as follows. At $T < T_c$, there exists a condensate of the Cooper pairs. This condensate is superfluid. It means that the dissipation-free electric current is carried by the Cooper pairs, i.e., the charge of an elementary current carrier is $2e$.

The microscopic theory of superconductivity was elaborated further by L.P. Gorkov (1958) who developed a method to solve the model BCS problem using Green's functions. He applied this method, in particular, to find microscopic interpretations for all phenomenological parameters of the GL theory as well as to define the theory's range of validity (see Sect. 6.6). The works by Gorkov have completed the development of the Ginzburg–Landau–Abrikosov–Gorkov theory (the GLAG theory).

Then everything seemed to be settled until the critical temperature made a quantum leap. In 1987 J. Bednorz and K.A. Müller discovered the first high-T_c superconductor ($LaBaCuO_4$, $T_c \sim 40$ K). Subsequently, materials have been found that raise T_c to temperatures of up to ~ 130 K ($HgBa_2Ca_2Cu_3O_8$). The classic BCS theory was unable to account for many of the properties of the high-T_c materials. The electron–phonon mechanism became questionable. New mechanisms – such as the so-called d-wave pairing – have been proposed. At present, the question of why the high-T_c superconductors have such high T_c values is still largely unanswered. Let us hope that the definitive answer need not be awaited for another 46 years.

Problems

Problem 1.1. A Pb specimen with a flat surface is placed in a magnetic field parallel to this surface. The field is equal to the critical magnetic field, $H_{cm}(T)$, and the temperature is 4.2 K. Find the surface current flowing within a 1 cm wide band.

Problem 1.2. Find the free energy gain associated with the superconducting transition for 1 cm^3 of Pb. The transition takes place at 4.2 K.

Problem 1.3. Under the conditions of Problem 1.2, find the amount of latent heat given up by 1 cm^3 of Pb during the transition into the superconducting state.

Problem 1.4. Find the temperature at which the heat capacities of Pb in the normal and superconducting states are equal.

Problem 1.5. Find the value of the thermodynamic critical field of Sn at $T = 3$ K.

Problem 1.6. Find the value of the heat capacity jump at the critical temperature for 1 cm^3 of Pb.

2. Linear Electrodynamics of Superconductors

2.1 The London Equations

In order to understand the behavior of a superconductor in an external electromagnetic field, let us use the so-called two-fluid model. We assume that all free electrons of the superconductor are divided into two groups: superconducting electrons of density n_s and normal electrons of density n_n. The total density of free electrons is $n = n_s + n_n$. As the temperature increases from 0 to T_c, the density n_s decreases from n to 0.

Let us start our systematic study of the superconductor in an electromagnetic field with the simplest case. Assume that both the electric and magnetic fields are so weak that they do not have any appreciable influence on the superconducting electron density. Assume, in addition, that the density n_s is the same everywhere, i.e., spatial variations of n_s are disregarded. The relation between the current, electric field, and magnetic field in this case is linear and described by the London equations [17].

2.1.1 The First London Equation

The equation of motion for superconducting electrons in an electric field is

$$n_s m \frac{\mathrm{d}\boldsymbol{v}_s}{\mathrm{d}t} = n_s e \boldsymbol{E} , \tag{2.1}$$

where m is the electron mass, e is the electron charge, $\boldsymbol{v}_s$ is the superfluid velocity, and n_s is the number density of the superfluid. Taking into account that the supercurrent density is $\boldsymbol{j}_s = n_s e \boldsymbol{v}_s$, we have

$$\boldsymbol{E} = \frac{\mathrm{d}}{\mathrm{d}t}(\Lambda \boldsymbol{j}_s) \tag{2.2}$$

with

$$\Lambda = m/n_s e^2 . \tag{2.3}$$

Equation (2.2) is simply Newton's second law for the superconducting electrons. It follows from this equation that in the stationary state, that is, when $\mathrm{d}\boldsymbol{j}_s/\mathrm{d}t = 0$, there is no electric field inside the superconductor. Note that here we disregard all possible spatial variations of the chemical potential of the superconducting electrons. Such variations become important, for example, when one considers the interface between a superconducting and a normal region, with a non-zero current passing through the interface. More details about this situation can be found in Chap. 7.

2.1.2 The Second London Equation

Let us now work out the relation between the supercurrent and the magnetic field in a superconductor. We denote the true microscopic magnetic field at a given point of a superconductor by $\boldsymbol{H}(\boldsymbol{r})$. Here an additional explanation is needed. We remember from Chap. 1 that magnetic field does not penetrate the interior of a type-I superconductor. Now we shall find that this is true only to a certain extent. Magnetic field does in fact penetrate to a shallow depth (of the order of 500–1000 Å) at the surface of a body. Our task at the moment is to work out how this field, $\boldsymbol{H}(\boldsymbol{r})$, varies in space.

Assume that the free energy of a superconductor in zero magnetic field is $\mathcal{F}_{s0}$. The kinetic energy density of the supercurrent is

$$W_{\text{kin}} = \frac{n_s m v_s^2}{2} = \frac{m j_s^2}{2 n_s e^2} . \tag{2.4}$$

Taking into account Maxwell's equation

$$\text{curl}\, \boldsymbol{H} = \frac{4\pi}{c} \boldsymbol{j}_s , \tag{2.5}$$

expression (2.4) for W_{kin} becomes

$$W_{\text{kin}} = \frac{\lambda^2}{8\pi} (\text{curl}\, \boldsymbol{H})^2 , \tag{2.6}$$

where we have defined

$$\lambda^2 = \frac{mc^2}{4\pi n_s e^2} . \tag{2.7}$$

As we already know, the magnetic energy density at some point of the superconductor is $H^2/8\pi$. Therefore, the free energy of the superconductor as a whole, including both the kinetic energy of the supercurrent and the magnetic field energy, is

$$\mathcal{F}_{sH} = \mathcal{F}_{s0} + \frac{1}{8\pi} \int [\boldsymbol{H}^2 + \lambda^2 (\text{curl}\, \boldsymbol{H})^2] \, \mathrm{d}V . \tag{2.8}$$

The integration is carried out over the entire volume of the superconductor.

Let us now solve a variational problem, namely, let us find the identity of the function $\boldsymbol{H}(\boldsymbol{r})$ corresponding to the minimum value of $\mathcal{F}_{sH}$. We note that a more accurate approach would be to minimize the Gibbs free energy. We shall indeed do that when deriving the Ginzburg–Landau equations (see Sect. 3.2). However, the result is independent of which particular function is examined for a minimum, $\mathcal{F}_{sH}$ or $\mathcal{G}_{sH}$. Therefore, we examine $\mathcal{F}_{sH}$ as it is easier to do.

Let us ascribe a small variation $\delta\boldsymbol{H}(\boldsymbol{r})$ to the function $\boldsymbol{H}(\boldsymbol{r})$. The resulting change in $\mathcal{F}_{sH}$ will be $\delta\mathcal{F}_{sH}$:

$$\delta\mathcal{F}_{sH} = \frac{1}{8\pi}\int(2\boldsymbol{H}\cdot\delta\boldsymbol{H} + 2\lambda^2 \mathrm{curl}\,\boldsymbol{H}\cdot\mathrm{curl}\,\delta\boldsymbol{H})\,\mathrm{d}V\,. \tag{2.9}$$

The function $\boldsymbol{H}(\boldsymbol{r})$ to be found is the function that makes $\mathcal{F}_{sH}$ a minimum, that is,

$$\delta\mathcal{F}_{sH} = 0\,. \tag{2.10}$$

Using the identity

$$\boldsymbol{a}\cdot\mathrm{curl}\,\boldsymbol{b} = \boldsymbol{b}\cdot\mathrm{curl}\,\boldsymbol{a} - \mathrm{div}\,[\boldsymbol{a}\times\boldsymbol{b}] \tag{2.11}$$

and combining (2.9) and (2.10), we obtain

$$\int[\boldsymbol{H} + \lambda^2(\mathrm{curl}\,\mathrm{curl}\,\boldsymbol{H})]\cdot\delta\boldsymbol{H}\,\mathrm{d}V - \int\mathrm{div}\,[\mathrm{curl}\,\boldsymbol{H}\times\delta\boldsymbol{H}]\,\mathrm{d}V = 0\,. \tag{2.12}$$

The second integral in (2.12) is zero, as one can see from the following. By Gauss's theorem, the second term in (2.12) can be written in the form $\oint[\mathrm{curl}\,\boldsymbol{H}\times\delta\boldsymbol{H}]\cdot\mathrm{d}\boldsymbol{S}$, where the integration is carried out over the surface of the superconductor. But the field at the surface is fixed, it is the external field; therefore $\delta\boldsymbol{H}(\boldsymbol{r}) = 0$ there.

We have arrived at the equation $\int(\boldsymbol{H} + \lambda^2\mathrm{curl}\,\mathrm{curl}\,\boldsymbol{H})\cdot\delta\boldsymbol{H}\,\mathrm{d}V = 0$. For an arbitrary variation $\delta\boldsymbol{H}(\boldsymbol{r})$, this equation can be satisfied only if the sum in parentheses is zero. Thus we have obtained the equation for the magnetic field in a superconductor:

$$\boldsymbol{H} + \lambda^2\mathrm{curl}\,\mathrm{curl}\,\boldsymbol{H} = 0\,. \tag{2.13}$$

This is the second London equation. It can also be written in a different form. Using Maxwell's equation (2.5) and the equality $\boldsymbol{H} = \mathrm{curl}\,\boldsymbol{A}$, we obtain from (2.13)

$$\boldsymbol{j}_s = -\frac{c}{4\pi\lambda^2}\boldsymbol{A}\,. \tag{2.14}$$

One can go from (2.13) to (2.14) only on the condition that the so-called London gauge is chosen for the vector potential:

$$\mathrm{div}\,\boldsymbol{A} = 0\,, \tag{2.15}$$

$$\boldsymbol{A}\cdot\boldsymbol{n} = 0\,, \tag{2.16}$$

where $\boldsymbol{n}$ is the unit vector normal to the superconductor's surface.

Equation (2.15), together with (2.14), specifies the conditions of continuity of the current and absence of a supercurrent source, while (2.16) assures that no supercurrent can pass through the boundary of a superconducting body.

It is assumed, of course, that there are no external circuits or contacts to current leads.

Using (2.3) and (2.7), we can write (2.14) in the form

$$\boldsymbol{j}_s = -\frac{1}{c\Lambda}\boldsymbol{A}\,, \tag{2.17}$$

$$\Lambda = 4\pi\lambda^2/c^2\,. \tag{2.18}$$

In the rest of the book, we shall often use the second London equation in the form (2.17).

2.2 Magnetic Field Penetration Depth

Let us apply the London equations to examine how a magnetic field penetrates a superconductor. Consider a superconducting semispace $x > 0$. The surface of the superconductor coincides with the plane $x = 0$. An external magnetic field H_0 is oriented along the z axis. To solve this problem, we shall use (2.13). Taking into account the symmetry of the problem and that $\operatorname{curl}\operatorname{curl}\boldsymbol{H} = -\nabla^2\boldsymbol{H}$, we can rewrite (2.13) in the form

$$\mathrm{d}^2H/\mathrm{d}x^2 - \lambda^{-2}H = 0\,. \tag{2.19}$$

The boundary conditions are: $H(0) = H_0$, $H(\infty) = 0$. The second boundary condition takes account of the Meissner–Ochsenfeld effect. The solution of (2.19) is

$$H = H_0\,\mathrm{e}^{-x/\lambda}\,. \tag{2.20}$$

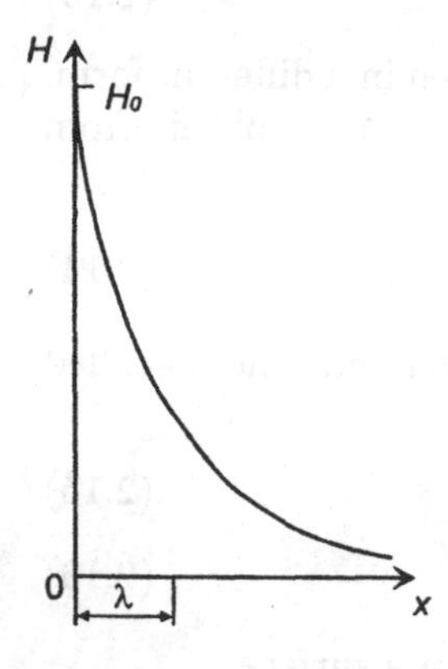

Fig. 2.1. Magnetic field penetration into a bulk superconductor. The field at the surface is H_0

It means that the magnetic field falls off with increasing distance from the surface of the superconductor. The characteristic decay length is λ (see

Fig. 2.1). This length clarifies the physical significance of the quantity λ formally defined by (2.7) and is called the London magnetic field penetration depth:

$$\lambda = \left(\frac{mc^2}{4\pi n_s e^2}\right)^{1/2} . \tag{2.21}$$

It follows that the screening (Meissner) supercurrent at the surface falls off over the same length. Indeed, the supercurrent is $\boldsymbol{j}_s = (c/4\pi)\,\mathrm{curl}\,\boldsymbol{H}$, which in our simple geometry reduces to $j_s = (c/4\pi)\,\mathrm{d}H/\mathrm{d}x$. Substituting (2.20) in the last expression, we get

$$j_s = \frac{cH_0}{4\pi\lambda}\,\mathrm{e}^{-x/\lambda} . \tag{2.22}$$

One can see from (2.21) that λ is temperature-dependent because it depends on n_s. A good approximation for the temperature dependence of λ is given by the empirical formula

$$\lambda(T) = \frac{\lambda(0)}{\left[1-(T/T_c)^4\right]^{1/2}} . \tag{2.23}$$

Let us now evaluate $\lambda(0)$. At $T = 0$ all electrons are superconducting, that is, $n_s = n \approx 10^{22}\ \mathrm{cm}^{-3}$. Substituting this in (2.21), together with $m \sim 10^{-27}$ g, $c \sim 3 \times 10^{-10}\ \mathrm{cm\,s}^{-1}$, $e = 4.8 \times 10^{-10}$ esu, we obtain $\lambda(0) \sim 600$ Å.

The values of $\lambda(0)$ for some superconductors are given in Table 2.1.

Table 2.1. London penetration depths for some superconductors [2]

Element	Al	Cd	Hg	In	Nb	Pb	Sn	Tl	$YBa_2Cu_3O_7$
$\lambda(0)$ /Å	500	1300	380–450 (anisotropy)	640	470	390	510	920	1700

2.3 Nonlocal Electrodynamics of Superconductors

Everything said so far about the electrodynamics of superconductors falls into the category of the so-called local electrodynamics. For instance, London's equation (2.17) relates the supercurrent density $\boldsymbol{j}_s$ (that is, the velocity of the supercurrent carriers, $\boldsymbol{v}_s$) to the vector potential $\boldsymbol{A}$ at the *same point.* Therefore, strictly speaking, (2.17) is applicable only if the size of the current carriers is much smaller than the characteristic length over which the vector potential changes, that is, smaller than the penetration depth λ. We know that the superconducting current carriers are pairs of electrons. Let us denote the size of a pair by ξ_0. For pure metals, ξ_0 is of the order $\xi_0 \sim 10^{-4}$ cm, as we shall find out later on, in Chap. 6. On the other hand, the penetration depth is $\lambda \sim 10^{-5}$–10^{-6} cm. Therefore, the local London electrodynamics is not applicable to pure superconductors, because the magnetic field changes appreciably over the length ξ_0.

Hence, the local equation (2.17) must be replaced by a nonlocal one relating the velocity of a particle to the magnetic field which is allowed to change substantially over the size of the particle, ξ_0. Such a nonlocal relation was proposed by A.B. Pippard [18] several years before the microscopic theory of superconductivity appeared.

In its general form, the nonlocal relation between $\boldsymbol{j}_s$ and $\boldsymbol{A}$ can be written as

$$\boldsymbol{j}_s(\boldsymbol{r}) = \int \hat{Q}(\boldsymbol{r}-\boldsymbol{r}')\,\boldsymbol{A}(\boldsymbol{r}')\cdot \mathrm{d}\boldsymbol{r}' , \tag{2.24}$$

where $\hat{Q}$ is an operator which, operating on the vector $\boldsymbol{A}$, converts it into the vector $\hat{Q}\boldsymbol{A}$. The operation range of the operator $\hat{Q}(\boldsymbol{r}-\boldsymbol{r}')$ is taken to be ξ_0, that is, $\hat{Q}(\boldsymbol{r}-\boldsymbol{r}')$ is nonzero only for $|\boldsymbol{r}-\boldsymbol{r}'| \le \xi_0$. This is how the effect of the vector $\boldsymbol{A}$ on a large particle (a supercurrent carrier) is averaged out. If we now let the operation range go asymptotically to zero, $\hat{Q}$ turns into the δ-function, and we come back to local electrodynamics.

Pippard's proposal was to choose $\hat{Q}\boldsymbol{A}$ in the form

$$\hat{Q}(\boldsymbol{r}-\boldsymbol{r}')\,\boldsymbol{A}(\boldsymbol{r}') = -\frac{3n_s e^2}{4\pi mc\xi_0}\frac{(\boldsymbol{r}-\boldsymbol{r}')}{(\boldsymbol{r}-\boldsymbol{r}')^4}\left[\boldsymbol{A}(\boldsymbol{r}')\cdot(\boldsymbol{r}-\boldsymbol{r}')\right]\mathrm{e}^{-|\boldsymbol{r}-\boldsymbol{r}'|/\xi_0} . \tag{2.25}$$

The exact form of the magnetic field penetration into a superconductor in the nonlocal case differs from the exponential dependence. However, in this case one can also speak of a magnetic field penetration depth defined as

$$\lambda = \frac{1}{H_0}\int_0^\infty H\,\mathrm{d}x . \tag{2.26}$$

Here H_0 is the field at the surface of a semi-infinite superconductor. If the field near the surface falls off exponentially, all three definitions of λ [(2.26), (2.19), and (2.20)] coincide.

We shall not attempt here to solve the nonlocal problem explicitly. Rather, we shall show how to find an estimate of the correct answer [4]. Assume that the true dependence $H(x)$ is approximated by an exponential dependence with a new penetration depth denoted by λ_P (Pippard's penetration depth). Then, the vector potential $\boldsymbol{A}$ acts on a particle of size ξ_0 only within the depth $\lambda_\mathrm{P} \ll \xi_0$ near the surface of the particle. The result is that, although the particle still participates in creating the current density $\boldsymbol{j}_s$, the effect of $\boldsymbol{A}$ on it is reduced, because only the part of it λ_P/ξ_0 'feels' the presence of the vector potential. Accordingly, the current density is reduced by a factor of ξ_0/λ_P in comparison with the local case. Inserting this coefficient in (2.14) we obtain

$$j_s = -\frac{c}{4\pi\lambda^2}\frac{\lambda_P}{\xi_0}A\,. \tag{2.27}$$

Rewriting (2.27) in the form

$$j_s = -\frac{c}{4\pi\lambda_P^2}A\,, \tag{2.28}$$

we obtain, just as we wished, the exponential penetration of the magnetic field to the depth λ_P. Comparing (2.27) and (2.28), we get $\lambda_P^2 = \lambda^2\xi_0/\lambda_P$, which leads to the following estimate for λ_P:

$$\lambda_P \approx (\lambda^2\xi_0)^{1/3}\,. \tag{2.29}$$

The quantity λ in (2.29) is defined by (2.7). Then it follows from (2.29) that, in the case $\lambda \ll \xi_0$, we have $\lambda_P \gg \lambda$, i.e., the nonlocal electrodynamics predicts deeper penetration of the magnetic field than would be expected from the London equations. This statement assumes, of course, that λ_P also satisfies the inequality $\lambda_P \ll \xi_0$, which is not always the case even for pure metals.

A typical representative of superconductors which are well-described by the nonlocal relations (the so-called Pippard superconductors) is Al. In contrast, Pb, even of high purity, is a London superconductor. When the temperature T approaches T_c, all superconductors become local (London superconductors) because λ diverges at $T \to T_c$ while ξ_0 is independent of temperature.

Everything said so far applies to pure metals, that is, those characterized by a mean free path $l \gg \xi_0$. If a metal contains a large number of impurities, $l \ll \xi_0$ can occur. We shall refer to such metals as dirty superconductors. Alloys also fall into this category. In very dirty metals, the role of the coherence length is played by the mean free path l. With the help of the microscopic theory, one can show that the magnetic field penetration depth for dirty superconductors is $\lambda_d \approx \lambda(\xi_0/l)^{1/2}$ at $l \ll \xi_0$. Thus, superconducting alloys are described well by the local London equations. In the rest of the book we shall use local equations only.

2.4 Quantum Generalization of the London Equations. Magnetic Flux Quantization

2.4.1 Generalized London Equation

It was already mentioned in Chap. 1 that the elementary carrier of the supercurrent is a pair of electrons called the Cooper pair. All pairs occupy the same energy level, or the same quantum state, i.e., they form a condensate. The wavefunction of a particle in the condensate can be written in the form

$\Psi(\boldsymbol{r}) = (n_s/2)^{1/2} e^{i\theta(\boldsymbol{r})}$, where θ is the phase of the wavefunction. The normalization of $\Psi(\boldsymbol{r})$ has taken into account that the density of electron pairs is $n_s/2$, where n_s is the density of the superconducting electrons.

Consider a particle of mass $2m$ and charge $2e$ moving in a magnetic field. Let us show that its momentum can be written as

$$\hbar\nabla\theta = 2m\boldsymbol{v}_s + \frac{2e}{c}\boldsymbol{A}, \tag{2.30}$$

where $\hbar$ is Planck's constant. In the absence of magnetic field, the particle flow density $n_s\boldsymbol{v}_s/2$ can be written in the form $(i\hbar/4m)(\Psi\nabla\Psi^* - \Psi^*\nabla\Psi)$. Substituting here the expression $\Psi(\boldsymbol{r}) = (n_s/2)^{1/2} e^{i\theta}$, we get $\hbar\nabla\theta = 2m\boldsymbol{v}_s$. Thus the total momentum of a particle moving in a magnetic field, $\hbar\nabla\theta$, is a sum of the momentum $2m\boldsymbol{v}$ and the momentum $(2e/c)\,\boldsymbol{A}$ due to the magnetic field.

Taking the expression for the supercurrent density in the form

$$\boldsymbol{j}_s = n_s e \boldsymbol{v}_s \tag{2.31}$$

and using (2.7) and (2.18), it is easy to obtain from (2.30) the generalized second London equation:

$$\boldsymbol{j}_s = \frac{1}{c\Lambda}\left(\frac{\Phi_0}{2\pi}\nabla\theta - \boldsymbol{A}\right). \tag{2.32}$$

Here we use the notation $\Phi_0 = \pi\hbar c/e$. The quantity Φ_0 has the dimensions of magnetic flux. Its physical significance will become clear in the following section.

2.4.2 Magnetic Flux Quantization

We now proceed to a remarkably interesting phenomenon: magnetic flux quantization in superconductors.

Consider a hole through a bulk superconductor, such as shown in Fig. 2.2. At first, the temperature is $T > T_c$ and the superconductor is in the normal state. Then we apply an external field H_0 parallel to the axis of the cylinder, and lower the temperature so that the specimen goes into the superconducting state. The field is now pushed out of the interior of the superconductor, while in the hole some frozen magnetic flux remains. This flux is produced by the supercurrent generated at the internal surface of the hole. Let us find this frozen magnetic flux.

Consider the contour C inside the superconductor, as in Fig. 2.2, enclosing the hole so that the distance between the contour and the internal surface of the hole is everywhere well in excess of λ. Then at any point of the contour, the supercurrent is $\boldsymbol{j}_s = 0$ and the path integral of (2.32) along the contour reduces to

$$\frac{\Phi_0}{2\pi}\oint_C \nabla\theta\cdot d\boldsymbol{l} = \oint_C \boldsymbol{A}\cdot d\boldsymbol{l}. \tag{2.33}$$

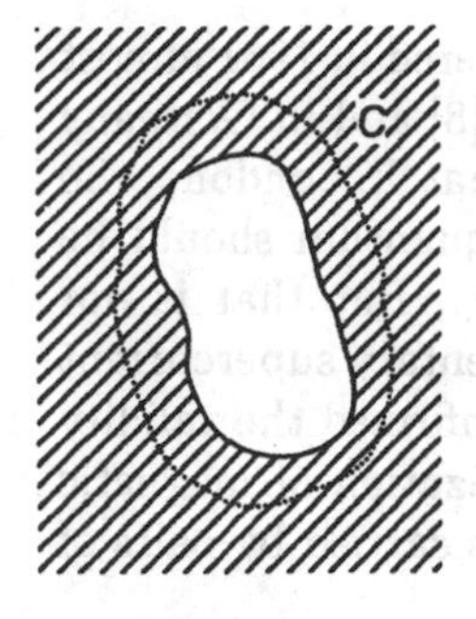

Fig. 2.2. Solid superconductor (*dashed area*) contains a hole. The contour C goes around the hole through the interior of the superconductor

Taking into account that

$$\oint_C \boldsymbol{A} \cdot \mathrm{d}\boldsymbol{l} = \Phi \,, \tag{2.34}$$

we have

$$\Phi = \frac{\Phi_0}{2\pi} \oint_C \nabla\theta \cdot \mathrm{d}\boldsymbol{l} \,. \tag{2.35}$$

Here Φ is the total magnetic flux through the contour C. From (2.35), one can see immediately that θ is a multiple-valued function; it changes by a certain value after every full circle around the hole. But the wavefunction Ψ must be single-valued. Therefore, we have to stipulate that the change in θ after a full circle around the hole containing the magnetic flux must be an integral multiple of $2\pi n$, where $n = 0, 1, 2 \ldots$. Indeed, the addition of $2\pi n$ to $\theta(\boldsymbol{r})$ does not change the function $\Psi(\boldsymbol{r}) = (n_s/2)^{1/2}\,\mathrm{e}^{\mathrm{i}\theta}$, because $\mathrm{e}^{\mathrm{i}2\pi n} = 1$. Therefore, $\oint_C \nabla\theta \cdot \mathrm{d}\boldsymbol{l} = 2\pi n$ and (2.35) can finally be written as

$$\Phi = n\Phi_0 \,, \tag{2.36}$$

where

$$\Phi_0 = \frac{\pi\hbar c}{e} = \frac{hc}{2e} \,. \tag{2.37}$$

It follows from (2.36) that the magnetic flux enclosed in the hole (or, more accurately, the magnetic flux through the contour C) can only assume values that are integral multiples of the minimum possible value of magnetic flux, the magnetic flux quantum Φ_0. The value of Φ_0 is defined by (2.37):

$$\Phi_0 = 2.07 \times 10^{-7}\ \mathrm{G\,cm^2} \,.$$

Physically, the origin of the magnetic flux quantization is the same as the quantization of electron orbits in atom. The wavefunction of electrons moving along a closed orbit must contain an integral number of wavelengths over the length of the orbit.

Experimentally, magnetic flux quantization was discovered in 1961 almost simultaneously in the USA (B. Deaver and W. Fairbank) [8] and in Germany (R. Doll and M. Näbauer) [9]. It is interesting to note that F. London, who predicted the flux quantization, believed that the flux quantum should be hc/e, that is, he predicted a value twice as large as Φ_0. But that is not surprising because he believed that the charge of an elementary supercurrent carrier was equal to the electron charge e. Experiment confirmed the validity of (2.37). Thus the experiments that detected the quantization of magnetic flux also provided direct evidence that the supercurrent is carried by pairs of electrons.

2.5 Magnetic Field and Current Distributions in Simple Configurations of Superconductors

2.5.1 Thin Slab in a Parallel Magnetic Field

In this section we shall analyze magnetic field and current distributions in simple configurations of superconductors. We start with the case of an infinite slab of thickness d placed in a uniform magnetic field H_0 parallel to its surface. Assume that the plane $x = 0$ is in the center of the slab and that its surfaces coincide with the planes $x = \pm d/2$. The magnetic field is along the z axis.

In the interior of the slab, the magnetic field H must satisfy (2.13). Then, from symmetry arguments, H is along the z axis and depends only on the x coordinate. Therefore (2.13) can be rewritten as:

$$\mathrm{d}^2H/\mathrm{d}x^2 - \lambda^{-2}H = 0 \tag{2.38}$$

with the boundary conditions $H(\pm d/2) = H_0$. The general solution of (2.38) is

$$H = H_1 \cosh(x/\lambda) + H_2 \sinh(x/\lambda)\,, \tag{2.39}$$

where H_1 and H_2 are integration constants. Substituting the boundary conditions into (2.39), we get two algebraic equations with two unknowns that can be solved easily. The final result is

$$H(x) = H_0 \frac{\cosh(x/\lambda)}{\cosh(d/2\lambda)}\,. \tag{2.40}$$

The supercurrent density in the slab can be found by combining (2.40) and Maxwell's equation $\mathrm{curl}\,\boldsymbol{H} = (4\pi/c)\boldsymbol{j}_\mathrm{s}$:

$$j_\mathrm{s} = -\frac{c}{4\pi}\frac{\mathrm{d}H}{\mathrm{d}x}\,. \tag{2.41}$$

As a result, we get

$$j_\mathrm{s} = -\frac{cH_0}{4\pi\lambda}\frac{\sinh(x/\lambda)}{\cosh(d/2\lambda)}\,. \tag{2.42}$$

If the slab is thick ($d \gg \lambda$), it follows from (2.40) and (2.42) that both the magnetic field and the current penetrate into it only to a certain depth of the order λ. In the other limiting case, a thin film ($d \ll \lambda$), we get, in linear approximation and after expanding the hyperbolic functions in terms of the small parameters x/λ and $d/2\lambda$:

$$H = H_0 \,, \quad j_s = \frac{cH_0 x}{4\pi\lambda^2} \,.$$

This means that the magnetic field penetrates the entire film, while the supercurrent density varies linearly with depth.

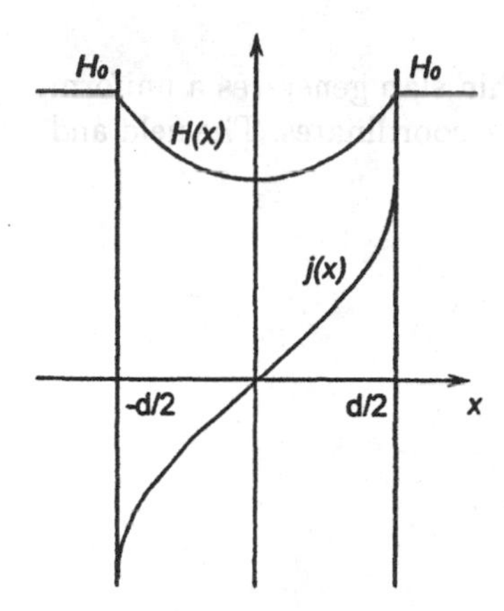

Fig. 2.3. Distribution of current and magnetic field across a thin film placed in a parallel uniform magnetic field

The current and field distributions obtained above are shown in Fig. 2.3. The current circulates at the edges of the slab so that the induced magnetic field cancels out the external field H_0 in its interior.

2.5.2 Thin Slab Carrying a Current

In this section we consider an infinite thin slab carrying a current in the absence of magnetic field. We assume that the slab is the same as in Sect. 2.5.1 and the current flows in the direction of the y axis. Further we assume that the current is uniform along the z axis, that is, edge effects are disregarded.

Consider a part of the slab in the form of a strip of unit width parallel to the z axis. The current I flows along the z axis and generates a magnetic field $\pm H_I$ at the surfaces of the slab ($x = \pm d/2$). Substituting these boundary conditions into the general solution (2.39), we find the field distribution in the slab:

$$H(x) = -H_I \frac{\sinh(x/\lambda)}{\sinh(d/2\lambda)} \,, \tag{2.43}$$

where $H_I = 2\pi I/c$.

Using Maxwell's equation (2.41), we find the distribution of current:

$$j_s(x) = \frac{cH_I}{4\pi\lambda} \frac{\cosh(x/\lambda)}{\sinh(d/2\lambda)} . \tag{2.44}$$

It follows from (2.43) and (2.44) that, similarly to the previous case, for $d \gg \lambda$ both the magnetic field and the current are present only within a surface layer of thickness λ. If the film is thin ($d \ll \lambda$), the pattern is different: the current flows through its entire cross-section and the field is a linear function of the coordinate:

$$H = -H_I \frac{2x}{d} , \qquad j_s = \frac{cH_I}{2\pi d} = \frac{I}{d} .$$

Recall that a uniform current in an infinite thin slab generates a uniform magnetic field outside the slab, independent of the coordinates. The field and current distributions are shown in Fig. 2.4.

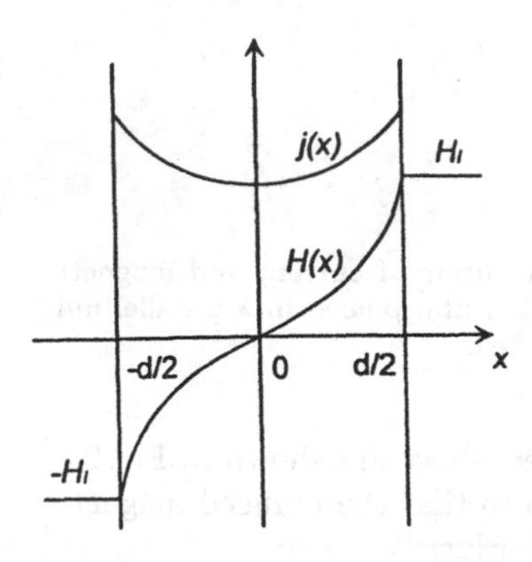

Fig. 2.4. Distribution of current and magnetic field across a thin film carrying a current

2.5.3 Thin Slab Carrying a Current in a Transverse Uniform Field

Consider a thin slab in a uniform external field H_0 parallel to the z axis. The slab carries a current along the y direction which is uniform along the z axis, as in Sect. 2.5.2. The total current through the cross-section of unit height is I. It generates a field $\pm H_I$ at the surfaces of the slab ($x = \pm d/2$). Thus, in this problem we have a superposition of the conditions from the two previous problems. Due to the linearity of the London equations, this should lead to a superposition of their solutions.

Consider a particular case $H_I = H_0$. The external field H_0 in this case cancels out the field generated by the current on one side of the slab and doubles it on the other side. As a result, the current I flows on one side of the slab only. Such a situation can be realized if the external field, H_0, is generated by a second slab carrying the same current I but in the opposite direction,

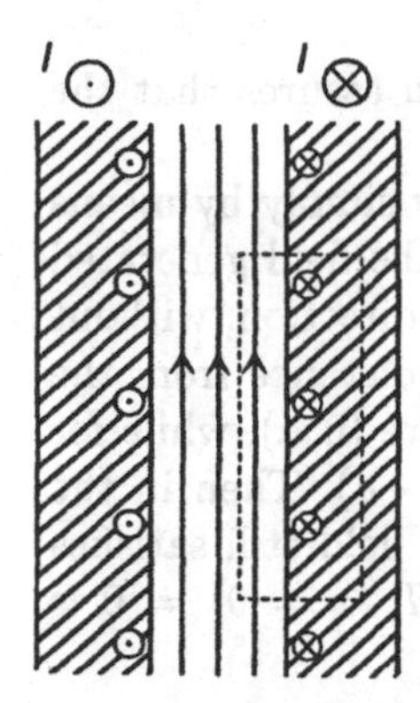

Fig. 2.5. If two parallel superconducting slabs carry currents which are equal in value and opposite in direction, the resulting magnetic field is 'locked' in the gap between the slabs

as in Fig. 2.5. The field in the gap between the slabs[1] is $H = (4\pi/c)\,I$. One can assume that this field is generated by both currents. Due to the current I flowing in the left-hand slab, there is a uniform field $H_I = 2\pi I/c$ directed upwards everywhere to the right of the slab and downwards everywhere to the left of it. Vice versa, the current from the right-hand slab generates a field H_I directed downwards everywhere to the right of the slab and upwards everywhere to the left of it. The result is that both to the right and to the left of the pair of slabs the fields cancel each other out while in the gap the field is doubled.

2.5.4 Thin Film Above a Superconducting Semispace

In this section we discuss a situation which is particularly important from a practical point of view: a thin film above a semi-infinite superconducting space (we shall refer to the latter as a superconducting screen). In order to find the distribution of the magnetic field when the film carries a certain current, we start with a very simple case.

Consider a rectilinear current-carrying conductor, which is placed above a superconducting screen and oriented along the y axis. Its distance from the screen is a. Let us find the magnetic field above the screen. If there were no screen, the magnetic field lines would form concentric circles centered on the conductor. But because the field lines cannot penetrate a superconductor (due to the Meissner–Ochsenfeld effect), it is obvious that, with the superconducting screen in place, the magnetic field around it will be distorted. Let us find this field.

In the semispace $z > 0$ and outside the conductor, there are no electric currents and $\operatorname{curl} \boldsymbol{H} = 0$. Therefore, we can introduce a magnetic field potential satisfying the Laplace equation. On the other hand, we know that the magnetic field at the surface of a superconductor is always tangential to the

[1] Stokes' theorem states that $\oint \boldsymbol{H} \cdot \mathrm{d}\boldsymbol{l} = (4\pi/c)\,I$, where I is the total current through the surface bounded by the contour along which the path integral is taken. Carrying out the integration along the dashed contour (Fig. 2.5), we obtain $H = (4\pi/c)\,I$.

surface, that is, $H_z(z = 0) = 0$. This boundary condition assures that the solution of the Laplace equation is single-valued.

The correct field in the region $z > 0$ can be found very simply by means of the method of images. It is easy to see that it will be the field generated by two rectilinear currents, equal in value and opposite in direction, without the superconducting screen. For one of the currents, the distance from the plane $z = 0$ is a, i.e., the coordinates of its cross-section are $(0, a)$, while for the other (the image of the first) the coordinates are $(0, -a)$. Then in the semispace $z > 0$ and outside the conductor, the magnetic field still satisfies the equation $\operatorname{curl} \boldsymbol{H} = 0$, and the boundary condition $H_z(z = 0) = 0$ is satisfied automatically due to symmetry of the problem.

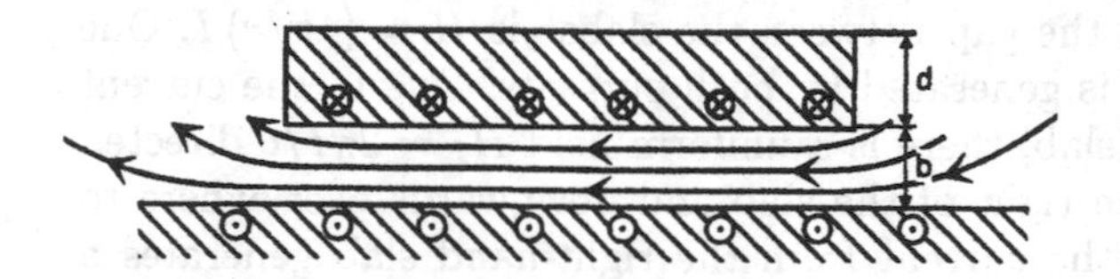

Fig. 2.6. Magnetic field generated by a current-carrying film placed above a superconducting semi-space

Thus we can apply the method of images to derive the field generated by a superconducting film placed above a semi-infinite superconducting screen. Let the thickness of the film be $d > \lambda$ and its width $w \gg \lambda$. The film is placed at a distance b from the screen and carries a current I. Let us find the current distributions in the film and in the screen and the magnetic field in the gap between the film and the screen.

Using the method of images, we replace the effect of the screen with that of an image film placed at a distance $2b$ from our film. The image also carries a current, but the direction of this current is opposite to that in the real film. Thus we have arrived at the problem of field and current distributions for two parallel films with opposite currents. This problem has just been solved in Sect. 2.5.3 (Fig. 2.5) and we already know the answer: in the gap between the film and the screen, there is a uniform magnetic field $H_I = \frac{4\pi}{c}\frac{I}{w}$. The current I in the film flows only near the surface adjacent to the screen, within a layer $\sim \lambda$. The current in the screen also flows within a layer $\sim \lambda$ near the surface and its surface density is I/w. The direction of this current is opposite to that of the current in the film. The relation between the field H_I and the current in the screen is given by the law of the total current. This situation is shown schematically in Fig. 2.6. The edge effects are disregarded in the above consideration. The magnetic field distribution incorporating the edge effects is shown in Fig. 2.7 [19].

The results that we have just obtained agree with reality if the real film playing the role of the screen can indeed be approximated by a semi-infinite screen. Below there is an example of a situation in which such an approximation is not valid.

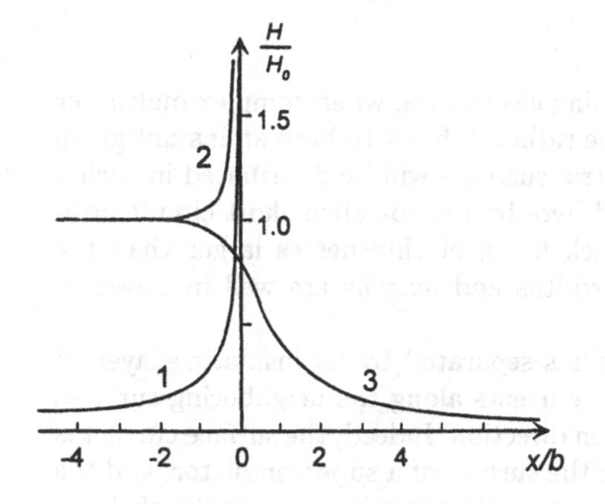

Fig. 2.7. Magnetic field distribution near the edge of a current-carrying superconducting film placed in the vicinity of a superconducting screen: *1* – field at the upper surface of the film; *2* – field at the lower surface of the screen; *3* – field at the surface of the screen

Consider a thick superconducting film (of thickness $d \geq \lambda$) of width w and length l deposited onto a glass substrate; $w \ll \lambda$. The film has been covered with a thin insulating layer of thickness b, such that $b \ll w$, and then another superconducting film has been deposited on top, also of thickness w and of length well in excess of l. Then a current from an external source is applied to the upper superconducting film. It is easy to realize that the

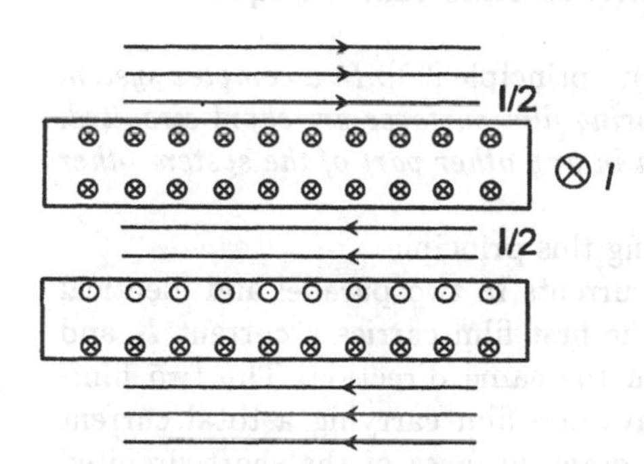

Fig. 2.8. Field and current distributions for two parallel closely spaced films. A current I is applied to the upper film. The lower film is not supplied with current

lower film cannot produce any screening effect in this situation. Indeed, if the upper film carries a certain current I, this current generates a magnetic field in the gap between the films. This means that the currents at the lower surface of the upper film and at the upper surface of the lower film must be equal in value. Furthermore, because the length of the lower film is finite, the current at its upper surface has to close at its lower surface, as in Fig. 2.8. As a result, the magnetic field at this surface is exactly the same as that in the gap between the films. Moreover, the current in the upper film is now split equally between its upper and lower surfaces, as illustrated in Fig. 2.8. Thus, the magnetic fields above and below the pair of films and in the gap are exactly the same as those in the absence of the lower film. That is, the latter does not produce any screening effect.

2.5.5 Short-Circuit Principle

Quite often in modern superconducting electronics, when complex multilayer systems are being designed, it can be rather difficult to have an instant grasp of how the magnetic fields and electric currents will be distributed in such a system. Great help can be provided here by the so-called short-circuit principle [20]. It can be applied to thick films, of thicknesses larger than the penetration depth, provided their widths and lengths are well in excess of the distances between the films.

Consider two superconducting films separated by an insulating layer. If the films carry electric currents, the currents along the neighboring surfaces must be equal in value and opposite in direction. Indeed, the surface current is determined by the magnetic field at the surface of a superconductor, and the magnetic field in the gap between the two films is common to both of them. If the thickness of the insulating layer is small, the magnetic flux through this layer is also small and the magnetic field at the edges of the films, associated with this flux, is negligible. Hence we can ignore the magnetic field at the edges and argue that the other parts of the films 'are not aware' of the existence of this flux and 'will not notice' if it disappears. And it can disappear as a result of short-circuiting, that is, bringing into contact two neighboring superconducting surfaces carrying electric currents that are equal in value and opposite in direction.

We can now formulate the short-circuit principle [20]: *In a complex system of superconducting films, if two neighboring film surfaces are short-circuited, it will not affect the current distribution in any other part of the system other than the two short-circuited surfaces.*

Consider several examples illustrating this principle.

(1) Let us find the distribution of currents in two parallel and identical thick films separated by a thin gap. The first film carries a current I_1 and the second one carries a current I_2, in the same direction. The two films are short-circuited. As a result, we have one film carrying a total current $I_1 + I_2$ distributed uniformly over the outer surfaces of the short-circuited films. Thus, the current at each of those surfaces is $(I_1 + I_2)/2$. We now know the currents at the outer surfaces of our system of films. Let i denote the current at the inner surfaces of the films. Because the total current in the first film, I_1, is fixed, we have $(I_1 + I_2)/2 + i = I_1$, i.e., $i = (I_1 - I_2)/2$. All surface currents in our system are now defined.

Note that if $I_2 = 0$, we come straight to the result shown in Fig. 2.8, that is, to the absence of screening by a superconducting film.

(2) If an insulated current-carrying wire is covered with a superconducting sheath, the latter will not be capable of screening the magnetic field generated by the current in the wire. To elucidate this statement, let us short-circuit the inner surface of the sheath and the surface of the wire. Then the entire current from the wire flows at the outer surface of the sheath and generates a magnetic field in the surrounding space.

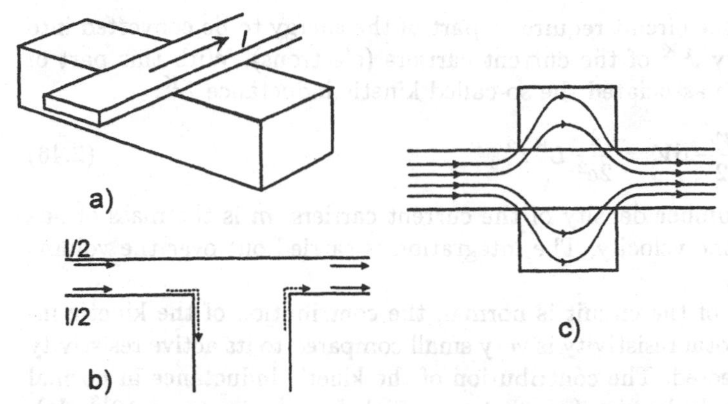

Fig. 2.9. Current-carrying film separated from the flat surface of a bulk superconductor by a thin gap: (a) general view; (b) side view after short-circuiting; (c) top view

(3) Let us find the distribution of currents in a system consisting of a thick superconducting film carrying a current which is placed above a massive superconducting bar (see Fig. 2.9). If the lower surface of the film and the upper surface of the bar are short-circuited, the resulting distribution of currents is that sketched in Fig. 2.9 (b). Indeed, at first the current from the film spreads over the upper surface of the bar, but a small fraction of it goes around the bar, along its lower surface. As a result, the current density at the upper surface of the film decreases sharply. Taking this into account, one can argue that in the initial system, without the short-circuit, the current in the part of the film immediately above the bar flows predominantly at the lower surface of the film. A current of the same value and opposite in direction must then appear at the upper surface of the bar, in the part situated immediately below the film. Further, this current spreads over the upper surface of the bar and closes there, while a small fraction of it closes over a path around the bar.

One can conclude that the role of the bar in this system approaches the role of the semi-infinite screen (see Fig. 2.7).

2.6 Kinetic Inductance

The inductance of a section of an electric circuit is usually defined by the energy $\mathcal{F}^{\mathrm{M}}$ of the magnetic field generated by a current I through the circuit:

$$\mathcal{F}^{\mathrm{M}} = \frac{1}{8\pi}\int H^2\,\mathrm{d}V = \frac{1}{2c^2}\,L^{\mathrm{M}} I^2\,. \tag{2.45}$$

The integral is taken over the entire space. We shall refer to this inductance as a magnetic, or geometrical, inductance. In addition, the process of generating the current I in the circuit requires a part of the energy to be converted into the kinetic energy $\mathcal{F}^{\mathrm{K}}$ of the current carriers (electrons). With this part of the energy can be associated the so-called kinetic inductance L^{K}:

$$\mathcal{F}^{\mathrm{K}} = \int n \frac{mv^2}{2} \mathrm{d}V = \frac{1}{2c^2} L^{\mathrm{K}} I^2 , \tag{2.46}$$

where n is the number density of the current carriers, m is the mass of one carrier and v is the velocity. The integration is carried out over the volume of the conductor.

If the section of the circuit is normal, the contribution of the kinetic inductance to the total resistivity is very small compared to its active resistivity and usually neglected. The contribution of the kinetic inductance in normal conductors can only be significant at very high frequencies (over 10^{13} Hz). In superconductors, however, the kinetic inductance sometimes plays an important role.

Recall that the supercurrent density is $\boldsymbol{j}_{\mathrm{s}} = n_{\mathrm{s}} e \boldsymbol{v}_{\mathrm{s}}$. Then, from (2.46), we obtain the following definition of the kinetic inductance of a superconductor:

$$L^{\mathrm{K}} = c^2 \Lambda \int \frac{{j_{\mathrm{s}}}^2}{I^2} \mathrm{d}V , \tag{2.47}$$

where the integral is taken over the volume of the superconductor and I is the total current through the superconductor.

Let us illustrate the concept of kinetic inductance with specific examples.

(1) Consider a superconducting wire of length l and radius R. Assume that $R \gg \lambda$. If the wire carries a current I, this current must circulate near its surface. The current density j_{s} at a distance r from the center of the wire is $j_{\mathrm{s}}(x) = j_{\mathrm{s}0}\, \mathrm{e}^{-x/\lambda}$, where $x = R - r$, $j_{\mathrm{s}0} = j_{\mathrm{s}}(0)$. The total current is $I = 2\pi R \lambda j_{\mathrm{s}0}$. Substituting these into (2.47) and carrying out the integration, we get $L^{\mathrm{K}} = l\lambda/R$.

Let us now introduce another quantity which is particularly useful in many applications: the inductance per square $L_{\square}$. For a flat superconductor, the greater its length, the larger its inductance and resistivity. Furthermore, the greater its width, the smaller the inductance. Therefore, the inductance of a square is always the same for a given superconductor, whether one considers 1 km^2 or 1 mm^2 of it. For our wire, since the circumference of the cross-section is $2\pi R$, the kinetic inductance per square is

$$L_{\square}^{\mathrm{K}} = 2\pi\lambda \, . \tag{2.48}$$

If we use (2.45) and calculate the part of the magnetic inductance associated with the magnetic field penetrating the superconductor (more accurately, penetrating the layer of the order λ near the surface) we obtain the same result:

$$L_{\square}^{\mathrm{M}} = 2\pi\lambda \, . \tag{2.49}$$

This result is also valid for the flat surface of a superconducting semispace.

The total inductance per square of the layer λ at the surface of a bulk superconductor is equal to the sum of (2.48) and (2.49):

$$L_\square = 4\pi\lambda\,. \tag{2.50}$$

In the International System of Units (SI), equation (2.50) takes the form: $L_\square = \mu_0\lambda$, with $\mu_0 = 4\pi \times 10^{-7}$ H/m. As follows from (2.48–50), in CGS $L_\square$ is measured in cm, and in SI in H, where 1 cm corresponds to 10^{-9} H = 1 nH.

One can say that λ is a characteristic of the inertia of the current carriers since, for a semispace, $L^{\mathrm{K}}_\square$ depends only on the penetration depth. If $\lambda \approx 5 \times 10^{-6}$ cm, then $L_\square = 4\pi\lambda = 6.3 \times 10^{-5}$ cm $= 6.3 \times 10^{-14}$ H.

(2) Consider the kinetic inductance of a thin superconducting film. Assume that the thickness of the film is $d \ll \lambda$ and, therefore, the current in the film is distributed uniformly over its thickness. We shall consider a small section of width w along the width of the film and assume that the current there is also uniform. Restricting the length of the section to the same value w, we obtain for the kinetic energy of the superconducting electrons within this section:

$$\mathcal{F}^{\mathrm{K}}_\square = \frac{\Lambda}{2} {j_s}^2 w^2 d\,.$$

Recalling that the current I through the cross-section is assumed to be uniform, we have $j_s = I/wd$ and

$$\mathcal{F}^{\mathrm{K}}_\square = \frac{1}{2d}\Lambda I^2 = \frac{1}{2d}\frac{4\pi\lambda^2}{c^2}I^2 = \frac{1}{2c^2}L^{\mathrm{K}}_\square I^2\,.$$

From the last expression we immediately obtain $L^{\mathrm{K}}_\square$ for a thin film:

$$L^{\mathrm{K}}_\square = 4\pi\lambda^2/d\,. \tag{2.51}$$

Now it is obvious that, for $d \ll \lambda$, the kinetic inductance can be significant. For example, for a thin film ($d \sim 10^{-6}$ cm) with the penetration depth $\lambda = 3 \times 10^{-5}$ cm, the kinetic inductance per square is, according to (2.51), $L^{\mathrm{K}}_\square \approx 10^{-2}$ cm $= 10^{-11}$ H.

(3) Finally, consider a thick film above a massive superconducting screen. Assume that the distance between the film and the surface of the screen is b. If the film carries a current, there must be a magnetic field present in the gap between the film and the screen. After calculating the energy of this magnetic field, we find its contribution to the inductance of the film $L_\square = 4\pi b$. In addition, the magnetic field penetrates the film to the depth λ_1 and the screen to the depth λ_2. This penetrating field, from (2.50), makes an additional contribution to the inductance of the system as a whole so that its total inductance is

$$L_\square = 4\pi(b + \lambda_1 + \lambda_2)\,. \tag{2.52}$$

One can see from (2.52) that, in order to reduce the inductance, the film should be placed as close to the screen as possible. However, making b substantially less than λ_1 or λ_2 is not of much use because both the magnetic and the kinetic inductance will remain within a layer of the order of the penetration depth, in the film as well as in the screen.

2.7 Complex Conductivity of a Superconductor

This section deals with the complex conductivity of a superconductor in an electromagnetic field. In what follows we assume that the electron mean free path l is small and, therefore, that the normal skin effect approximation is valid. In other words, l is small enough and the frequencies are low enough that l is less than the penetration depth of the electromagnetic field. The frequency of electron collisions is $\tau^{-1} = v_F/l \gg \omega$, where ω is the frequency of the electromagnetic wave and v_F is the electron velocity on the Fermi surface.

In the following analysis we shall partly follow the monograph by Van Duzer and Turner [21].

In order to calculate the conductivity of a superconductor in a high-frequency field, we use the two-fluid model, that is, we assume that there are normal electrons of density n_n and superconducting electrons of density n_s, and the total density of conduction electrons is $n = n_s + n_n$. The motion of the superconducting electrons is governed by the first London equation (2.2):

$$\boldsymbol{E} = \Lambda \, \mathrm{d}\boldsymbol{j}_s/\mathrm{d}t \,. \tag{2.53}$$

For the normal electrons we can write

$$e\boldsymbol{E} - \frac{m}{n_n e}\frac{\boldsymbol{j}_n}{\tau} = \frac{m}{n_n e}\frac{\mathrm{d}\boldsymbol{j}_n}{\mathrm{d}t} \,. \tag{2.54}$$

The left-hand side of (2.54) describes the forces on the normal electrons: the electric field and the average 'friction' due to electron collisions. The right-hand side is the product of the electron mass and acceleration. For one normal electron, this Newton's second law can be written as

$$\boldsymbol{E} = \frac{n_s}{n_n}\Lambda\frac{\mathrm{d}\boldsymbol{j}_n}{\mathrm{d}t} + \frac{n_s}{n_n}\Lambda\frac{\boldsymbol{j}_n}{\tau} \,. \tag{2.55}$$

Assuming $\boldsymbol{j}_s \propto \mathrm{e}^{\mathrm{i}\omega t}$, we can rewrite (2.53) and (2.55) as

$$\boldsymbol{j}_s = -\mathrm{i}\frac{1}{\Lambda\omega}\boldsymbol{E} \,, \tag{2.56}$$

$$\boldsymbol{j}_n = \frac{n_n}{n_s}\frac{\tau}{\Lambda}\frac{1 - \mathrm{i}\omega\tau}{1 + (\omega\tau)^2}\boldsymbol{E} \,. \tag{2.57}$$

The total current density is $\boldsymbol{j} = \boldsymbol{j}_s + \boldsymbol{j}_n$, and we finally have

$$\boldsymbol{j} = \sigma\boldsymbol{E} \,, \qquad \sigma = \sigma_1 - \mathrm{i}\sigma_2 \,, \tag{2.58}$$

$$\sigma_1 = \frac{n_n}{n_s}\frac{\tau}{\Lambda}\frac{1}{1 + (\omega\tau)^2} \,, \tag{2.59}$$

$$\sigma_2 = \frac{1}{\Lambda\omega}\left[1 + \frac{n_\mathrm{n}}{n_\mathrm{s}}\,\frac{(\omega\tau)^2}{1+(\omega\tau)^2}\right] . \tag{2.60}$$

Equations (2.58–60) describe the complex conductivity of a superconductor in a high-frequency electromagnetic field.

2.8 Skin Effect and Surface Impedance

2.8.1 Normal Skin Effect

It is well known that an electromagnetic field penetrates a normal metal to the so-called skin depth, or to the depth of a skin layer. In this section we shall consider how the field penetrates a superconductor. The surface of the superconductor coincides with the plane $x = 0$.

Let us write down Maxwell's equations

$$\operatorname{curl} \boldsymbol{H} = \frac{4\pi}{c}\,\sigma \boldsymbol{E} , \tag{2.61}$$

$$\operatorname{curl} \boldsymbol{E} = -\frac{1}{c}\,\frac{\partial \boldsymbol{H}}{\partial t} . \tag{2.62}$$

Assuming that the magnetic field in the superconductor is $\boldsymbol{H} \propto \mathrm{e}^{-\mathrm{i}(kx-\omega t)}$ and taking the curl of both sides of (2.61), we get

$$-\nabla^2 \boldsymbol{H} = -\frac{4\pi}{c}\,\sigma\,\frac{\partial \boldsymbol{H}}{\partial t} . \tag{2.63}$$

Here we have used the equations $\operatorname{div} \boldsymbol{H} = 0$ and $\operatorname{curl}\operatorname{curl} \boldsymbol{H} = -\nabla^2 \boldsymbol{H}$. Substituting $\boldsymbol{H} \propto \mathrm{e}^{-\mathrm{i}(kx-\omega t)}$ in (2.63), we obtain

$$k^2 = -\mathrm{i}\,\frac{4\pi}{c^2}\,\sigma\omega , \tag{2.64}$$

and

$$k = (1-\mathrm{i})/\delta , \tag{2.65}$$

where

$$\delta = \left(\frac{c^2}{2\pi\sigma\omega}\right)^{1/2} . \tag{2.66}$$

Our problem is now essentially solved. Indeed, the field penetration is defined by the quantity k which can be expressed through the conductivity σ using (2.65) and (2.66). Let us make some simplifications. Assume that the temperature is not too close to T_c so that $(n_\mathrm{n}/n_\mathrm{s})\,(\omega\tau)^2 \ll 1$. In addition, we assume, as always, that $\omega\tau \ll 1$. Then we obtain from (2.59) and (2.60)

$$\sigma = \frac{n_\mathrm{n}}{n_\mathrm{s}}\,\frac{\tau}{\Lambda} - \mathrm{i}\,\frac{1}{\Lambda\omega} . \tag{2.67}$$

Substituting (2.67) into (2.66), we get

$$\delta = \frac{\sqrt{2}\lambda}{\left(\frac{n_n}{n_s}\omega\tau - \mathrm{i}\right)^{1/2}} . \tag{2.68}$$

At low frequencies, when $(n_n/n_s)(\omega\tau) \ll 1$, we have $\delta = \sqrt{2\mathrm{i}}\lambda = \lambda(1+\mathrm{i})$. Substituting this into (2.65), we obtain $k = -\mathrm{i}/\lambda$, i.e., $H \propto \mathrm{e}^{-ikx} = \mathrm{e}^{-x/\lambda}$. Thus, as one should have expected, a low-frequency magnetic field penetrates a superconductor in the same manner as a dc field, that is, it is exponentially damped over the magnetic field penetration depth. In the general case, the field penetration is described by (2.65) and (2.68).

2.8.2 Surface Impedance

By definition, the surface impedance is

$$Z = \frac{4\pi}{c}\frac{E}{H} . \tag{2.69}$$

This expression has a clear physical meaning. Suppose that there are ac electric and magnetic fields at the surface of a metal, such that the vectors $\boldsymbol{E}$ and $\boldsymbol{H}$ are mutually orthogonal and tangential to the surface. Then $c\boldsymbol{H}/4\pi$ is the density of the surface current $\boldsymbol{j}_{\text{surf}}$ and, therefore, Z in (2.69) is E/j_{surf}, that is, it expresses the surface impedance per square.

Let us now work out the surface impedance of a superconductor. With the magnetic field written as $\boldsymbol{H} \propto \mathrm{e}^{-\mathrm{i}(kx-\omega t)}$, equation (2.61) becomes

$$\mathrm{i}kH = \frac{4\pi}{c}\sigma E . \tag{2.70}$$

Then from (2.69) and (2.65) we have

$$Z\frac{\mathrm{i}k}{\sigma} = \frac{1+\mathrm{i}}{\sigma\delta} . \tag{2.71}$$

Substituting (2.66) and (2.67) into (2.71), we get

$$Z = R_\square + \mathrm{i}X_{L\square} , \tag{2.72}$$

$$R_\square = \frac{2\pi\omega^2\lambda}{c^2}\frac{n_n}{n_s}\tau , \tag{2.73}$$

$$X_{L\square} = \frac{4\pi\lambda\omega}{c^2} = \frac{\omega L_\square}{c^2} . \tag{2.74}$$

The real part of the impedance, $R_\square$, reflects the energy dissipation due to heating, while the imaginary part, $X_{L\square}$, is the inductive resistance.

Let us find the temperature dependences of $R_\square$ and $X_{L\square}$. Recalling the empirical relation $\lambda \propto (1-t^4)^{-1/2}$, we have $n_s \propto 1-t^4$, or $n_s = n(1-t^4)$

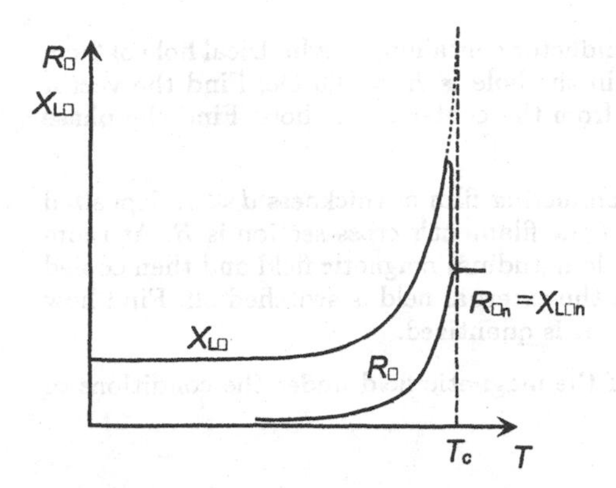

Fig. 2.10. Real and imaginary parts of the surface impedance as functions of temperature [22]

because $\lambda \propto n_s{}^{-1/2}$. Here $t = T/T_c$ and n is the density of free electrons in a metal. Then $n_n = nt^4$ and

$$R_\square \propto t^4/(1-t^4)^{3/2}\,, \qquad X_{L\square} \propto (1-t^4)^{-1/2}\,. \tag{2.75}$$

The above formulas reproduce the temperature dependences of both active and imaginary components of the impedance rather well (at least qualitatively), with an exception of temperatures close to T_c. In this temperature interval, (2.73) and (2.74) are no longer valid. Recall that, while deriving these formulas, we used (2.67) obtained on the assumption that $(n_n/n_s)\,(\omega\tau)^2 \ll 1$. But no matter how low the frequency is, the superconducting electron density at $T \to T_c$ is $n_s \to 0$ and the inequality is violated. Therefore, at $T \to T_c$ we have from (2.60): $\sigma = (n_n/n_s)\tau/\Lambda - \mathrm{i}\,(n_n/n_s)\,(\omega\tau)^2/\Lambda\omega$. Neglecting the imaginary part and substituting this expression in (2.65) and (2.71) we obtain

$$Z = \frac{2\pi}{c^2}\left(2\omega\tau\frac{n_s}{n_n}\right)^{1/2}\frac{\lambda}{\tau}(1+\mathrm{i}) = \frac{1+\mathrm{i}}{\sigma_n\delta_n}\,, \tag{2.76}$$

where σ_n and δ_n are the conductivity and the skin depth of the normal metal, respectively.

It follows from (2.76) that, at $T \to T_c$, $R_\square$ and $X_{L\square}$ are equal in value and independent of temperature, since $n_s{}^{1/2}\lambda = \text{const}$.

The temperature dependences of $R_\square$ and $X_{L\square}$ obtained within the two-fluid model are sketched in Fig. 2.10.

As far as high-temperature superconductors are concerned, many experiments indicate that they do not have a 'clean' gap. This means that even at zero temperature they have an intrinsic density of quasiparticle states in the gap down to zero energy. As a consequence, the temperature dependences of the two-fluid model are no longer valid for high-T_c materials.

Problems

Problem 2.1. Consider a bulk superconductor containing a cylindrical hole of 0.1 mm diameter. There are 7 magnetic flux quanta trapped in the hole. Find the magnetic field in the hole.

Problem 2.2. Consider a bulk superconductor containing a cylindrical hole of 2 cm diameter. The magnetic field trapped in the hole is $H = 300$ Oe. Find the vector potential $\boldsymbol{A}$ at the distance $R = 2$ cm from the center of the hole. Find the phase gradient $\nabla\theta$ at the same distance.

Problem 2.3. Consider a thin superconducting film of thickness $d \ll \lambda$ deposited onto a dielectric filament. The radius of the filament's cross-section is R. At room temperature, the filament is placed in a longitudinal magnetic field and then cooled down to a temperature below T_c. Then the external field is switched off. Find how the magnetic flux trapped by the filament is quantized.

Problem 2.4. Find the distribution of the magnetic field under the conditions of the previous problem.

3. The Ginzburg–Landau Theory

3.1 Introduction

The London theory (Chap. 2) did not take into account quantum effects. The first quantum (phenomenological) theory of superconductivity was the Ginzburg–Landau (GL) theory [23].

A quantum theory should take into account, firstly, that the superconducting state is more ordered than the normal one and, secondly, that the transition from one state to the other (without magnetic field) is a second-order phase transition. This implies the existence of an order parameter for a superconductor, which is nonzero at $T < T_c$ and vanishes at $T \geq T_c$. At the same time, in order to develop a quantum theory, it is necessary to introduce an effective wavefunction of the superconducting electrons, $\Psi(\boldsymbol{r})$.

Ginzburg and Landau decided to combine the two prerequisites: they decided to consider $\Psi(\boldsymbol{r})$ as an order parameter. This step required great scientific courage and deep physical intuition.

The GL theory is based on the theory of second-order phase transitions developed by Landau [24]. According to this theory, a phase transition of second order occurs when the state of a body changes gradually while its symmetry changes discontinuously at the transition temperature. Furthermore, the low-temperature phase is the one of reduced symmetry, i.e., it is more ordered. Such phenomena as the ferromagnetic transition at the Curie point, the transition of helium to the superfluid state, a number of order–disorder transitions in alloys, and the superconducting transition all fall into the category of second-order phase transitions.

Let us explain how a gradual change of the state of a body can be accompanied by a discontinuous change of its symmetry. It becomes apparent from the following example of a structural transition leading to a more ordered state. Consider a linear chain of atoms of two different types: A and B. Assume that at a sufficiently high temperature the probabilities that the atoms of type A and type B occupy one of the sites in the chain are equal (this state corresponds to complete disorder, as in Fig. 3.1 (a)). At lower temperatures, $T < T_c$, the probability that the atoms of type A occupy the various sites in the chain is sketched in Fig. 3.1 (b). Now the arrangement of the atoms is more ordered, that is, every other atom in the chain is of type A. Thus, on going down through T_c and to even lower temperatures, an order parameter

η appeared and gradually started to increase, while the period of the structure, which was a above T_c, underwent a discontinuous change at $T = T_c$ and became $2a$.

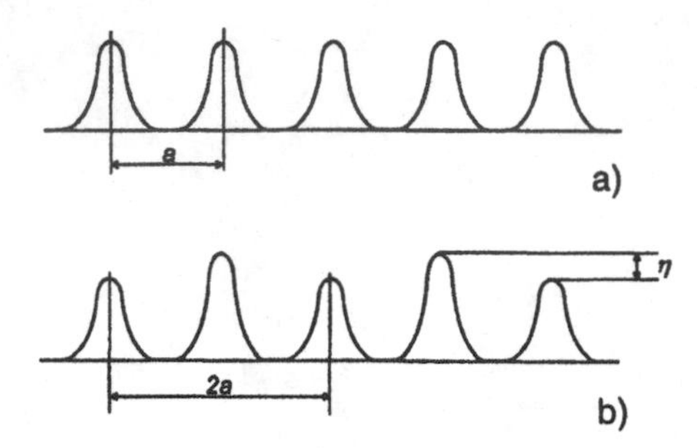

Fig. 3.1. Illustration of the discontinuous symmetry change in a structural phase transition of second order; **(a)** probability of finding an atom of type A at a given point of the crystal at $T > T_c$; **(b)** the same at $T < T_c$

Let us now go back to superconductivity. The Landau theory of second-order phase transitions is based on an expansion of the free energy in powers of the order parameter, which is small near the transition temperature. Such a starting point for the theory implies that it is valid only at temperatures close to the critical temperature, $T_c - T \ll T_c$. The range of validity of the GL theory will be discussed in more detail in Sect. 6.6.

Let us assume that the wavefunction of superconducting electrons $\Psi(\boldsymbol{r})$ is the order parameter. In addition, let us choose normalization of this wavefunction such that $|\Psi(\boldsymbol{r})|^2$ gives the density of Cooper pairs:

$$|\Psi(\boldsymbol{r})|^2 = n_s/2 . \tag{3.1}$$

Consider first the simplest case: a homogeneous superconductor without external magnetic field. In this case Ψ does not depend on $\boldsymbol{r}$ and the expansion of the free energy in powers of $|\Psi|^2$ near T_c becomes

$$F_{s0} = F_n + \alpha|\Psi|^2 + \frac{\beta}{2}|\Psi|^4 . \tag{3.2}$$

Here F_{s0} is the free energy density of the superconductor in the absence of magnetic field, F_n is its free energy density in the normal state and α and β are some phenomenological expansion coefficients which are characteristics of the material.

Let us find the value of $|\Psi|^2$ for which the free energy of a homogeneous superconductor, F_{s0}, is a minimum. This value, $|\Psi_0|^2$, is the solution of the equation

$$\frac{dF_{s0}}{d|\Psi|^2} = 0 .$$

Substituting here (3.2) and carrying out elementary calculations we obtain

$$|\Psi_0|^2 = -\alpha/\beta . \tag{3.3}$$

Substituting (3.3) into (3.2), we find the difference in energy

$$F_n - F_{s0} = \alpha^2/2\beta \,. \tag{3.4}$$

Recalling that, from (1.12), this difference equals $H_{cm}^2/8\pi$, we have

$$H_{cm}^2 = 4\pi\alpha^2/\beta \,. \tag{3.5}$$

Let us discuss the temperature dependence of the coefficients α and β. Since the order parameter must be zero at $T = T_c$ and finite at $T < T_c$, it follows from (3.3) that $\alpha = 0$ at $T = T_c$ and $\alpha < 0$ at $T < T_c$. Therefore we can write, to first order in $(T_c - T)$,

$$\alpha \propto (T - T_c) \,. \tag{3.6}$$

This temperature dependence of α correlates (3.5) with the empirical formula (1.1) near T_c.

The coefficient β is positive and independent of temperature. Indeed, as follows from (3.3), at $T < T_c$ and $\alpha < 0$, $|\Psi_0|^2$ can be positive only if $\beta > 0$. On the other hand, if $T > T_c$ (and consequently, from (3.6), $\alpha > 0$) and $\beta > 0$, the energy F_{s0} reaches its minimum at $|\Psi_0|^2 = 0$. This means that there is no superconducting state at $T > T_c$ as indeed ought to be the case. Thus we have $\beta > 0$ both at $T < T_c$ and at $T > T_c$. Therefore, to a first approximation in $(T_c - T)$, we may assume $\beta = \text{const}$.

3.2 Equations of the Ginzburg–Landau Theory

3.2.1 Free Energy Density

Let us now go over to the general case of an inhomogeneous superconductor in a uniform external magnetic field.

Near T_c, the Gibbs free energy can be expanded in powers of Ψ as

$$\begin{aligned} G_{sH} &= G_n + \alpha|\Psi|^2 + \frac{\beta}{2}|\Psi|^4 \\ &\quad + \frac{1}{2m^*}\left|-i\hbar\nabla\Psi - \frac{2e}{c}\boldsymbol{A}\Psi\right|^2 + \frac{H^2}{8\pi} - \frac{\boldsymbol{H}\cdot\boldsymbol{H}_0}{4\pi} \,, \end{aligned} \tag{3.7}$$

where G_n is the free energy density of a superconductor in the normal state and $\boldsymbol{H}_0$ is the external magnetic field. The last but one term in (3.7) simply represents the magnetic energy density, where $\boldsymbol{H}$ is the exact microscopic field at a given point of the superconductor. The term before that is the kinetic energy density of the superconducting electrons. Let us analyze the latter in more detail.

In quantum mechanics, the kinetic energy density of a particle of mass m is

$$\frac{1}{2m}|-i\hbar\nabla\Psi|^2 \,.$$

For a particle of charge e moving in the field with vector potential $\boldsymbol{A}$, the operator $-i\hbar\nabla$ in the expression for the kinetic energy density has to be modified:

$$-i\hbar\nabla \longrightarrow -i\hbar\nabla - \frac{e}{c}\boldsymbol{A} = m\boldsymbol{v} .$$

Therefore, the velocity operator is

$$\boldsymbol{v} = -(i\hbar/m)\,\nabla - (e/cm)\,\boldsymbol{A} .$$

Since it is the velocity $\boldsymbol{v}$ that enters the expression for the kinetic energy density, we can now understand why the corresponding term in (3.7) looks as it does. It should only be added that a substitution $e \longrightarrow 2e$ has been made in (3.7) which takes into account that the elementary charge carrier of the supercurrent is $2e$. Accordingly, m^* in (3.7) is twice the electron mass.

3.2.2 Ginzburg–Landau (GL) Equations

By (3.7), the Gibbs free energy of a superconductor as a whole is

$$\begin{aligned} \mathcal{G}_{sH} &= \mathcal{G}_n + \int \left[\alpha|\Psi|^2 + \frac{\beta}{2}|\Psi|^4 + \frac{1}{4m}\left| -i\hbar\nabla\Psi - \frac{2e}{c}\boldsymbol{A}\Psi \right|^2 \right. \\ &\quad \left. + \frac{(\operatorname{curl}\boldsymbol{A})^2}{8\pi} - \frac{(\operatorname{curl}\boldsymbol{A})\cdot\boldsymbol{H}_0}{4\pi} \right] \mathrm{d}V , \end{aligned} \tag{3.8}$$

where the integration is carried out over the entire volume of the superconductor. Our task now is to find equations for the functions $\Psi(\boldsymbol{r})$ and $\boldsymbol{A}(\boldsymbol{r})$ such that their solutions, when substituted in (3.8), give the minimum value of $\mathcal{G}_{sH}$.

In order to do that we shall first assume that $\Psi(\boldsymbol{r})$ and $\boldsymbol{A}(\boldsymbol{r})$ are invariant and then solve the variational problem with respect to $\Psi^*(\boldsymbol{r})$:

$$\delta_{\Psi^*}\mathcal{G}_{sH} = 0 , \tag{3.9}$$

$$\begin{aligned} \delta_{\Psi^*}\mathcal{G}_{sH} &= \int \mathrm{d}V \left[\alpha\Psi\,\delta\Psi^* + \beta\Psi\,|\Psi|^2\,\delta\Psi^* + \frac{1}{4m}\left(i\hbar\nabla\,\delta\Psi^* \right.\right. \\ &\quad \left.\left. - \frac{2e}{c}\boldsymbol{A}\,\delta\Psi^* \right)\cdot\left(-i\hbar\nabla\Psi - \frac{2e}{c}\boldsymbol{A}\Psi \right) \right] . \end{aligned} \tag{3.10}$$

The term $\delta\Psi^*$ could be taken out of the square brackets but for the term $i\hbar\nabla\,\delta\Psi^*$. Let us make some modifications. We write

$$\varphi = \left(-i\hbar\nabla\Psi - \frac{2e}{c}\boldsymbol{A}\Psi \right) .$$

Using the identity

$$\nabla\,(\delta\Psi^*\varphi) = \varphi\nabla\delta\Psi^* + \delta\Psi^*\nabla\varphi ,$$

we then have

$$\int \mathrm{d}V\,\nabla\delta\Psi^*\varphi = -\int \delta\Psi^*\nabla\varphi\,\mathrm{d}V + \int \nabla(\delta\Psi^*\varphi)\,\mathrm{d}V\,. \tag{3.11}$$

By Gauss's theorem, the last integral in (3.11) can be converted into a surface integral:

$$\int \nabla(\delta\Psi^*\varphi)\,\mathrm{d}V = \oint_S \delta\Psi^*\varphi\,\mathrm{d}S\,.$$

Substituting (3.11) into (3.10) and (3.10) into (3.9), we obtain

$$\begin{aligned}\delta_{\Psi^*}\mathcal{G}_{sH} &= \int \mathrm{d}V\left[\alpha\Psi + \beta\Psi|\Psi|^2 + \frac{1}{4m}\left(-\mathrm{i}\hbar\nabla - \frac{2e}{c}\boldsymbol{A}\right)^2\Psi\right]\delta\Psi^* \\ &\quad + \oint_S \left[-\mathrm{i}\hbar\nabla\Psi - \frac{2e}{c}\boldsymbol{A}\Psi\right]\delta\Psi^*\,\mathrm{d}S = 0\,.\end{aligned}$$

For an arbitrary function $\delta\Psi^*$, this expression can be zero only if both expressions in square brackets are zero. From this requirement we obtain the first equation of the GL theory and its boundary condition:

$$\alpha\Psi + \beta\Psi|\Psi|^2 + \frac{1}{4m}\left(\mathrm{i}\hbar\nabla + \frac{2e}{c}\boldsymbol{A}\right)^2\Psi = 0\,, \tag{3.12}$$

$$\left(\mathrm{i}\hbar\nabla\Psi + \frac{2e}{c}\boldsymbol{A}\Psi\right)\cdot\boldsymbol{n} = 0\,,$$

where $\boldsymbol{n}$ is the unit vector normal to the surface of the superconductor. One can easily verify that minimization of $\mathcal{G}_{sH}$ with respect to Ψ leads to the complex-conjugate of (3.12). Thus, we have obtained the equation for the order parameter Ψ. One variable still remains: $\boldsymbol{A}$. In order to obtain the equation for $\boldsymbol{A}$, we shall minimize the expression for $\mathcal{G}_{sH}$ (3.8) with respect to $\boldsymbol{A}$:

$$\begin{aligned}\delta_A\mathcal{G}_{sH} &= \int \mathrm{d}V\left\{\frac{1}{4m}\delta_A\left[\left(\mathrm{i}\hbar\nabla\Psi^* - \frac{2e}{c}\boldsymbol{A}\Psi^*\right)\cdot\left(-\mathrm{i}\hbar\nabla\Psi - \frac{2e}{c}\boldsymbol{A}\Psi\right)\right]\right. \\ &\quad \left. + \frac{1}{4\pi}\operatorname{curl}\boldsymbol{A}\cdot\operatorname{curl}\delta\boldsymbol{A} - \frac{\boldsymbol{H}_0}{4\pi}\cdot\operatorname{curl}\delta\boldsymbol{A}\right\} \\ &= \int\left\{\frac{1}{4m}\left(-\frac{2e}{c}\Psi^*\delta\boldsymbol{A}\right)\cdot\left(-\mathrm{i}\hbar\nabla\Psi - \frac{2e}{c}\boldsymbol{A}\Psi\right)\right. \\ &\quad + \frac{1}{4m}\left(\mathrm{i}\hbar\nabla\Psi^* - \frac{2e}{c}\boldsymbol{A}\Psi^*\right)\cdot\left(-\frac{2e}{c}\Psi\,\delta\boldsymbol{A}\right) \\ &\quad \left. + \frac{1}{4\pi}(\operatorname{curl}\boldsymbol{A} - \boldsymbol{H}_0)\cdot\operatorname{curl}\delta\boldsymbol{A}\right\}\mathrm{d}V\,. \end{aligned} \tag{3.13}$$

One notices that $\delta\boldsymbol{A}$ in (3.13) could be taken out of the brackets but for the term $(1/4\pi)(\operatorname{curl}\boldsymbol{A} - \boldsymbol{H}_0)\cdot\operatorname{curl}\delta\boldsymbol{A}$. Using the identity

$$\boldsymbol{a}\cdot\operatorname{curl}\boldsymbol{b} = \boldsymbol{b}\cdot\operatorname{curl}\boldsymbol{a} - \operatorname{div}[\boldsymbol{a}\times\boldsymbol{b}]\,, \tag{3.14}$$

we can carry out the integration in the last term of (3.13):

$$\frac{1}{4\pi}\int \mathrm{d}V\,(\mathrm{curl}\,\boldsymbol{A}-\boldsymbol{H}_0)\cdot\mathrm{curl}\,\delta\boldsymbol{A}=\frac{1}{4\pi}\int \mathrm{d}V\,\delta\boldsymbol{A}\cdot\mathrm{curl}\,\mathrm{curl}\,\boldsymbol{A}$$
$$-\frac{1}{4\pi}\oint \mathrm{d}\boldsymbol{S}\cdot[\delta\boldsymbol{A}\times(\mathrm{curl}\,\boldsymbol{A}-\boldsymbol{H}_0)]\,. \tag{3.15}$$

When doing so we have used Gauss's theorem to convert the volume integral into a surface integral:

$$\int \mathrm{d}V\,\mathrm{div}\,[\delta\boldsymbol{A}\times(\mathrm{curl}\,\boldsymbol{A}-\boldsymbol{H}_0)]=\oint \mathrm{d}\boldsymbol{S}\cdot[\delta\boldsymbol{A}\times(\mathrm{curl}\,\boldsymbol{A}-\boldsymbol{H}_0)]\,.$$

The surface integral is zero because the magnetic field at the surface of the superconductor is fixed; hence $\delta\boldsymbol{A}\,|_S=0$.

Substituting now (3.15), without the last term, into (3.13) and equating the free energy variation to zero, we get, after elementary modifications,

$$\begin{aligned}\delta_{\boldsymbol{A}}\mathcal{G}_{sH} &= \int\Bigg[\frac{\mathrm{i}\hbar e}{2mc}(\Psi^*\nabla\Psi-\Psi\nabla\Psi^*)\\ &\quad+\frac{2e^2}{mc^2}\boldsymbol{A}\,|\Psi|^2+\frac{1}{4\pi}\,\mathrm{curl}\,\mathrm{curl}\,\boldsymbol{A}\Bigg]\cdot\delta\boldsymbol{A}\,\mathrm{d}V=0\,.\end{aligned} \tag{3.16}$$

For arbitrary $\delta\boldsymbol{A}$, the integral in (3.16) can be zero only if the expression in square brackets is zero. This condition determines the second equation of the GL theory, for the vector potential $\boldsymbol{A}$:

$$\boldsymbol{j}_s=-\frac{\mathrm{i}\hbar e}{2m}(\Psi^*\nabla\Psi-\Psi\nabla\Psi^*)-\frac{2e^2}{mc}|\Psi|^2\boldsymbol{A}\,, \tag{3.17}$$

where, by Maxwell's equation, the current density $\boldsymbol{j}_s$ in the superconductor is

$$\boldsymbol{j}_s=\frac{c}{4\pi}\,\mathrm{curl}\,\mathrm{curl}\,\boldsymbol{A},\qquad \boldsymbol{H}=\mathrm{curl}\,\boldsymbol{A}\,. \tag{3.18}$$

Let us go over to a dimensionless wavefunction $\psi(\boldsymbol{r})$ by setting

$$\psi(\boldsymbol{r})=\Psi(\boldsymbol{r})/\Psi_0\,, \tag{3.19}$$

where $\Psi_0{}^2=n_s/2=|\alpha|/\beta$. In addition, we introduce two more definitions:

$$\xi^2=\frac{\hbar^2}{4m|\alpha|}\,, \tag{3.20}$$

$$\lambda^2=\frac{mc^2}{4\pi n_s e^2}=\frac{mc^2\beta}{8\pi e^2|\alpha|}\,. \tag{3.21}$$

Now the GL equations can be written in a more concise and convenient form:

$$\xi^2\left(\mathrm{i}\nabla+\frac{2\pi}{\Phi_0}\boldsymbol{A}\right)^2\psi-\psi+\psi\,|\psi|^2=0\,, \tag{3.22}$$

$$\mathrm{curl}\,\mathrm{curl}\,\boldsymbol{A}=-\mathrm{i}\frac{\Phi_0}{4\pi\lambda^2}(\psi^*\nabla\psi-\psi\nabla\psi^*)-\frac{|\psi|^2}{\lambda^2}\boldsymbol{A}\,. \tag{3.23}$$

Here $\Phi_0=\pi\hbar c/e$ is the flux quantum (see Sect. 2.4).

Furthermore, if ψ is written as $\psi = |\psi|\,\mathrm{e}^{\mathrm{i}\theta}$, the second GL equation becomes

$$\operatorname{curl}\operatorname{curl}\boldsymbol{A} = \frac{|\psi|^2}{\lambda^2}\left(\frac{\Phi_0}{2\pi}\nabla\theta - \boldsymbol{A}\right) . \tag{3.24}$$

From (3.12) we obtain a boundary condition for ψ. If the superconductor has an interface with vacuum or an insulator, the condition will be

$$\left(\mathrm{i}\nabla + \frac{2\pi}{\Phi_0}\boldsymbol{A}\right)\cdot \boldsymbol{n}\psi = 0 , \tag{3.25}$$

where $\boldsymbol{n}$ in the unit vector normal to the surface of the superconductor. It is easy with the help of (3.17) to verify that (3.25) satisfies a very natural physical requirement, that is, it assures that no supercurrent passes through the superconductor–dielectric interface. However, the same requirement (that the normal component of the supercurrent through the interface is zero) will also be satisfied if a more general condition is imposed:

$$\left(\mathrm{i}\nabla + \frac{2\pi}{\Phi_0}\boldsymbol{A}\right)\cdot \boldsymbol{n}\psi = \mathrm{i}a\psi , \tag{3.26}$$

where a is an arbitrary real number. With the help of the microscopic theory of superconductivity one can show that (3.26) is appropriate for the interface between a superconductor and a normal metal.

3.2.3 Gauge Invariance of the GL Theory

The GL equations contain the vector potential $\boldsymbol{A}$. But it is well-known that the choice of $\boldsymbol{A}$ is not unique. Indeed, the transformation

$$\boldsymbol{A} = \boldsymbol{A}' + \nabla\varphi , \tag{3.27}$$

where $\varphi(\boldsymbol{r})$ is a single-valued scalar function, does not change the magnetic field:

$$\boldsymbol{H} = \operatorname{curl}\boldsymbol{A} = \operatorname{curl}\boldsymbol{A}',$$

because

$$\operatorname{curl}\nabla\varphi = 0 .$$

To ensure that the theoretical results are independent of the choice of the vector potential, or, in other words, that they are gauge invariant, one has to impose the condition that the GL equations themselves are gauge invariant.

It is easy to verify that this condition is satisfied if we use the following transformations to go from the variables $\boldsymbol{A}$ and ψ to variables $\boldsymbol{A}'$ and ψ':

$$\boldsymbol{A} = \boldsymbol{A}' + \nabla\varphi , \tag{3.28}$$

$$\psi = \psi' \exp\left[\mathrm{i}\,\frac{2\pi}{\Phi_0}\varphi(\boldsymbol{r})\right] . \tag{3.29}$$

Let us verify this for (3.24). We proceed from $\boldsymbol{A}$ and ψ to $\boldsymbol{A}'$ and ψ' following (3.28) and (3.29). Then we see immediately that the second GL equation remains unaffected:

$$\operatorname{curl}\operatorname{curl}\boldsymbol{A}' = \frac{|\psi'|^2}{\lambda^2}\left(\frac{\Phi_0}{2\pi}\nabla\theta' - \boldsymbol{A}'\right).$$

Similarly, one can verify the gauge invariance of the first GL equation (3.22).

The gauge invariance of the GL equations implies a conclusion which is very important for our future discussions: *For a simply connected superconductor, it is always possible to choose a gauge for the vector potential* $\boldsymbol{A}$ *such that the function* $\psi(\boldsymbol{r})$ *is real.* (Recall that a simply connected body is a body in the interior of which an arbitrary closed contour can be reduced to a point without having to cross the boundaries of the body. In other words, there are no unbroken holes in a simply connected body. A toroid is an example of a doubly connected body.)

The requirement that a superconductor be simply connected is essential because, in a multiply connected superconductor, the phase of the order parameter, θ, is no longer a single-valued function. On going around a cavity in a multiply connected superconductor, the phase can vary by an integral multiple of 2π. Therefore, $\theta(\boldsymbol{r})$ becomes unsuitable as the corresponding gauge for $\boldsymbol{A}$.

3.3 Two Characteristic Lengths in Superconductors. The Proximity Effect

3.3.1 The Coherence Length and the Penetration Depth

In the preceding section, we formally introduced a quantity ξ in (3.20). Now we shall find its physical significance.

Let us consider a simple example. A thin film of a normal metal is deposited onto a clean flat surface of a superconductor. This will cause a local reduction in the superconducting electron density near the surface of the superconductor. In other words, the order parameter $|\psi|$ at the surface will be somewhat different from its equilibrium value deep inside the superconductor, where $|\psi| = 1$. What then is the characteristic length over which the order parameter returns to unity?

Let us take the x axis perpendicular to the surface of the superconductor ($x = 0$ at the surface). Then it is obvious that ψ can vary only along the x axis, i.e., $\psi = \psi(x)$. In addition, ψ can be presumed real because we are dealing with a simply connected superconductor (see Sect. 3.2). Then the first GL equation (3.22) reduces to a simple form:

$$-\xi^2\frac{\mathrm{d}^2\psi}{\mathrm{d}x^2} - \psi + \psi^3 = 0\,. \tag{3.30}$$

Suppose that the normal layer at the surface is so thin that the magnitude of ψ at the surface is not very different from 1, that is,

$$\psi = 1 - \varepsilon(x), \quad \varepsilon(x) \ll 1 .$$

Substituting this into (3.30) and keeping only linear terms in $\varepsilon(x)$, we obtain

$$\xi^2 \frac{d^2\varepsilon(x)}{dx^2} - 2\varepsilon(x) = 0 . \tag{3.31}$$

Allowing for the fact that as $x \to \infty$ the order parameter $\psi \to 1$, we have $\varepsilon(\infty) = 0$. Then the obvious solution of (3.31) is

$$\varepsilon = \varepsilon(0)\, e^{-\sqrt{2}\, x/\xi} .$$

It follows that ξ is indeed the characteristic scale over which variations of the order parameter ψ occur. This length is called the coherence length.

Another quantity, λ, introduced in (3.21), is already known to us (Sect. 2.2). This is the penetration depth for a weak magnetic field. Both ξ and λ are temperature dependent:

$$\lambda^2 = \frac{mc^2\beta}{8\pi|\alpha|e^2} , \tag{3.32}$$

$$\xi^2 = \frac{\hbar^2}{4m|\alpha|} . \tag{3.33}$$

In the vicinity of T_c, we have $|\alpha| \propto (T_c - T)$. Therefore, in the temperature interval close to T_c

$$\lambda \propto (T_c - T)^{-1/2} , \quad \xi \propto (T_c - T)^{-1/2} . \tag{3.34}$$

In the entire temperature interval, a good approximation for $\lambda(T)$ is given by the empirical formula

$$\lambda(T) = \lambda(0)\,(1 - T/T_c)^{-1/4} . \tag{3.35}$$

The values of $\lambda(0)$ for a number of superconducting materials are given in Table 2.1. Using (3.3) and (3.1), one can verify that the penetration depth (3.32) is the same as that introduced earlier in (2.7).

Now we can introduce another important quantity, the GL parameter κ:

$$\kappa = \lambda/\xi . \tag{3.36}$$

Another expression for κ can be obtained from (3.32), (3.33) and (3.36):

$$\kappa = 2\sqrt{2}\frac{e}{\hbar c}\,\lambda^2 H_{\rm cm} . \tag{3.37}$$

An additional, very useful, formula follows from (3.37) and the expression for the flux quantum $\Phi_0 = \pi\hbar c/e$:

$$\sqrt{2}\, H_{\rm cm} = \frac{\Phi_0}{2\pi\lambda\xi} . \tag{3.38}$$

Before concluding this section, let us discuss the effect of magnetic field on the order parameter and the penetration depth.

Suppose that a superconducting specimen characterised by $\kappa \ll 1$ (or $\lambda \ll \xi$) occupies the semispace $x > 0$, with an external magnetic field along the z axis. Since by assumption $\lambda \ll \xi$, the magnetic field penetrates the specimen only to a small depth, $d \ll \xi$. This implies that the order parameter Ψ is affected by the magnetic field only over a small depth λ, while it can be expected to vary in a substantial way over a much larger length, ξ. Therefore, over the major part of the coherence length, the order parameter is 'not aware' of the existence of the magnetic field and is close to $\Psi_0 = (|\alpha|/\beta)^{1/2}$. Hence we can conclude that, for $\kappa \ll 1$, the effect of the magnetic field on the order parameter is insignificant.

It follows immediately that the magnetic field penetration depth, which depends on $|\Psi|^2$, is also only slightly affected by the magnetic field. Rigorous calculations show that, in the presence of an external magnetic field H_0, the amplitude of the order parameter ψ is reduced near the surface of the superconductor by [23]

$$\frac{1}{4\sqrt{2}} \kappa \frac{{H_0}^2}{H_{\mathrm{cm}}^2} .$$

For $\lambda \gg \xi$, the effect of an external magnetic field on the order parameter is much greater and brings about a number of qualitatively new and interesting effects. We shall discuss these in detail in the following sections.

3.3.2 The Proximity Effect

The role of the coherence length becomes particularly evident when we consider a normal metal N and a superconductor S in good contact with each other. The Cooper pairs can penetrate from S into N and 'live' there for some time. As a result, a thin layer near the NS interface becomes superconducting. On the other hand, the penetration of the Cooper pairs from S into N results in their density being reduced in S. That is, in the vicinity of the interface, the order parameter ψ in the superconductor becomes less than 1, even without magnetic field.

This is the so-called proximity effect.

Consider the simplest case. Suppose that two superconductors, with slightly different critical temperatures (T_{cn} and T_{cs}), are in good contact: $T_{\mathrm{cs}} > T_{\mathrm{cn}}$, $T_{\mathrm{cs}} - T_{\mathrm{cn}} \ll T_{\mathrm{cn}}$. The temperature of the specimen T is chosen to satisfy the inequality $T_{\mathrm{cn}} < T < T_{\mathrm{cs}}$, hence the material with the critical temperature T_{cn} is in the normal state. The interface between the two materials is flat and coincides with the plane $x = 0$. The superconductor occupies the semispace $x > 0$, and the normal metal $x < 0$.

The behavior of the order parameter in the S region ($x > 0$) can be determined by solving the first GL equation (3.12) of the form (3.30). The integration can be carried out exactly. The first integral is

$$-\xi^2\left(\frac{d\psi}{dx}\right)^2-\psi^2+\frac{1}{2}\psi^4=C\,, \tag{3.39}$$

where C is the integration constant. Since at $x\to\infty$ we have $(d\psi/dx)\to 0$ and $\psi\to 1$, the integration constant is $C=-1/2$. Substituting this into (3.39) and carrying out the integration, we obtain

$$\psi=\tanh\left[(x-x_0)/\sqrt{2}\xi\right]\,. \tag{3.40}$$

Here x_0 is the integration constant to be determined from the boundary condition at $x=0$. In our case this condition is

$$\frac{1}{\psi}\frac{d\psi}{dx}=\frac{1}{b}\,. \tag{3.41}$$

The value of b, in general, should be calculated using the microscopic theory. Its geometrical significance becomes evident from Fig. 3.2.

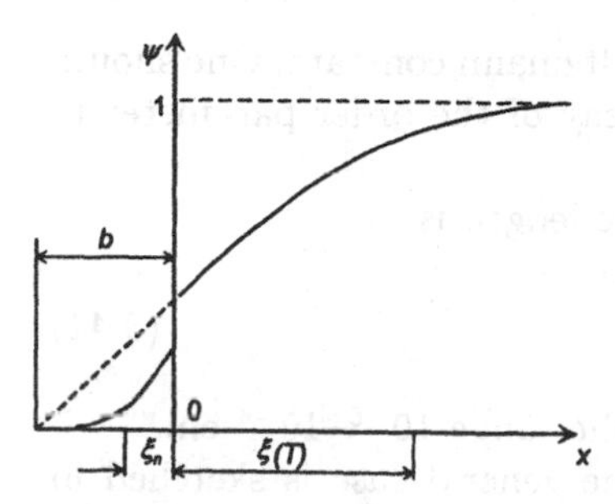

Fig. 3.2. Order parameter $\psi(x)$ near the interface between a superconductor ($x>0$) and a normal metal ($x<0$)

On substitution of (3.40) into (3.41), we find how the integration constant x_0 relates to b:

$$-\sinh\left(\sqrt{2}\,\frac{x_0}{\xi}\right)=\sqrt{2}\,\frac{b}{\xi}\,.$$

Consider now the behavior of the order parameter ψ in the normal region ($x<0$). The first GL equation is also applicable in this case (see, for example, [25]). Indeed, by (3.6), we have $\alpha_n\propto(T-T_{cn})$ so that $\alpha_n<0$ at $T<T_{cn}$ and $\alpha_n>0$ at $T>T_{cn}$. Hence, for the N region, the first GL equation (3.12) subject to the condition $T-T_{cn}\ll T_{cn}$ becomes

$$-\xi_n{}^2\frac{d^2\psi}{dx^2}+\psi+\psi^3=0\,,$$

where $\xi_n{}^2=\hbar^2/4m\alpha_n$.

The order parameter in the normal region is small ($\psi\ll 1$), therefore the cubic term can be neglected. Then

$$-\xi_n{}^2\frac{d^2\psi}{dx^2}+\psi=0\,.$$

Subject to the condition $\psi \to 0$ at $x \to \infty$, the solution is

$$\psi = \psi_0 \exp(-|x|/\xi_\mathrm{n}) \,. \tag{3.42}$$

From (3.42) we see that the order parameter penetrates the N region to the depth ξ_n while decaying exponentially over this distance. Since T_cn and T_cs are close to each other, ψ and $\mathrm{d}\psi/\mathrm{d}x$ can be assumed continuous at the NS interface. Then, from (3.42), we have $b = \xi_\mathrm{n}$.

In the general case of a superconductor in contact with a true normal metal ($T_\mathrm{cn} = 0$), the GL equations are not applicable to the normal region. Nevertheless, qualitatively, the phenomenon remains, that is, the order parameter penetrates the normal region to a certain depth ξ_n. Rigorous calculations based on the microscopic theory give the following results [26].

In a pure N metal, that is, when the electron mean free path is $l_\mathrm{n} \gg \xi_\mathrm{n}$, the coherence length is

$$\xi_\mathrm{n} = \frac{\hbar v_\mathrm{Fn}}{2\pi k_\mathrm{B} T} \,, \tag{3.43}$$

where v_Fn is the Fermi velocity and k_B is the Boltzmann constant. One should bear in mind, however, that at $T \to 0$ the decay of the order parameter in the N region is much slower than exponential.

In a 'dirty' N metal ($l_\mathrm{n} \ll \xi_\mathrm{n}$) the coherence length is

$$\xi_\mathrm{n} = \left(\frac{\hbar v_\mathrm{Fn} l_\mathrm{n}}{6\pi k_\mathrm{B} T}\right)^{1/2} \,. \tag{3.44}$$

Evaluations by (3.43,44) give values for ξ_n in the range 10^{-5}–10^{-4} cm.

The behavior of the order parameter in the general case is sketched in Fig. 3.2. According to [27], the value of b in the 'dirty' case is

$$b = \frac{\sigma_\mathrm{s}}{\alpha \sigma_\mathrm{n}} \xi_\mathrm{n} \,, \tag{3.45}$$

where σ_s and σ_n are the conductivities in the S and N regions, respectively, ξ_n is defined by (3.44), and the coefficient α is of the order 1. The exact values of α for a number of situations can be found in [27].

The existence of the proximity effect has been reliably established by experiment. If a superconducting film is deposited onto the surface of a normal metal, the critical temperature of the film decreases. Such a system was studied, for instance, in [28], namely a Pb film deposited on top of an Al film of thickness 4400 Å and $T_\mathrm{c} = 1.2\,\mathrm{K}$. When the thickness of the Pb film was $d\,(\mathrm{Pb}) = 900$ Å, the critical temperature of the whole system, T_c, was close to the critical temperature of bulk lead (7.2 K). However, as the Pb film was made thinner, $d\,(\mathrm{Pb}) = 600$ Å, T_c fell to approximately 5.6 K, and at $d\,(\mathrm{Pb}) = 200$ Å it became $T_\mathrm{c} = 1.6\,\mathrm{K}$, that is, close to the critical temperature of Al.

The proximity effect is utilized in SNS Josephson junctions where the phase coherence between the superconducting electrodes is established via a normal layer that can be quite thick ($\sim 10^{-4}$cm).

3.4 Energy of a Normal Metal–Superconductor Interface

Type-I and type-II superconductors, as we already know, can show entirely different responses to an external magnetic field. The reason is that the surface energy of the interface between a normal and a superconducting region, σ_{ns}, is positive for type-I superconductors and negative for those of type-II. We are now in a position to understand the origin of this difference. It turned out that for type-I superconductors $\lambda < \xi$, and for those of type-II $\lambda > \xi$. The dividing line between the superconductors of type-I and type-II will be determined more precisely later. Let us start with a type-I superconductor.

Consider a flat NS interface within a superconductor in the intermediate state. Assume that far away to the left of the interface, the material is superconducting, and far away to the right of the interface, it is normal. The interface is perpendicular to the x axis and the magnetic field is parallel to the z axis. Since we consider a superconducting semispace, i.e., a simply connected superconductor, we can always choose an appropriate gauge so that the wavefunction of the GL theory is real. In addition, due to the simple geometry of the problem, all variables depend only on x, and the vector potential $\boldsymbol{A}$ may be assumed parallel to the y axis. The origin $x = 0$ is chosen at the interface.

Thus the initial data are:

(1) $\boldsymbol{H} = (0,\ 0,\ H(x))$,
(2) $\psi = \psi(x)$ is a real function,
(3) $\boldsymbol{A} = (0,\ A(x),\ 0)$.

Then the GL equations (3.22) and (3.24) can be rewritten in the form

$$-\xi^2 \frac{\mathrm{d}^2\psi}{\mathrm{d}x^2} + \left(\frac{2\pi\xi}{\Phi_0}\right)^2 A^2\psi - \psi + \psi^3 = 0\,,$$
$$\frac{\mathrm{d}^2 A}{\mathrm{d}x^2} = \left(\frac{\psi^2}{\lambda^2}\right) A\,. \tag{3.46}$$

One can easily verify that the first integral of these equations is

$$\left[1 - \left(\frac{2\pi\xi A}{\Phi_0}\right)^2\right]\psi^2 - \frac{1}{2}\psi^4 + \left(\frac{2\pi\lambda\xi}{\Phi_0}\right)^2 \left(\frac{\mathrm{d}A}{\mathrm{d}x}\right)^2 + \xi^2\left(\frac{\mathrm{d}\psi}{\mathrm{d}x}\right)^2 = C\,, \tag{3.47}$$

where C is the integration constant which can be found from the boundary conditions: at $x \to -\infty$, $\psi \to 1$, $\mathrm{d}\psi/\mathrm{d}x \to 0$, and $A \to 0$.

Indeed, at the far left, that is, as $x \to -\infty$, there is no magnetic field and the wavefunction of the GL theory tends to 1. On substitution of these boundary conditions into (3.47), we obtain $C = 1/2$ and, using (3.38), we finally have

$$\left[\left(\frac{2\pi\xi A}{\Phi_0}\right)^2 - 1\right]\psi^2 + \frac{1}{2}\psi^4 = \xi^2\left(\frac{\mathrm{d}\psi}{\mathrm{d}x}\right)^2 + \frac{H^2}{2H_{cm}^2} - \frac{1}{2}\,. \tag{3.48}$$

Now, after these preliminary preparations, let us turn to calculation of the surface energy per unit area of the interface between a normal and a superconducting region. But first a little bit of physics. When inspecting a superconducting domain, we shall need to know the exact value of the external magnetic field surrounding it. As this domain is a part of the specimen in the intermediate state, there must be a normal domain next to it, which is penetrated by a magnetic field of strength H_{cm}. (Recall that this is the field strength established automatically in the normal regions of a superconductor in the intermediate state.)

Thus, with respect to the superconducting domain, the external field is always H_{cm}. Let us now write down the density of the Gibbs free energy deep in the interior of the superconducting domain, that is, far away to the left of the interface. From (1.22) we have

$$G_s = F_{s0} - HH_{cm}/4\pi \ .$$

Since at the far left $H = 0$, we get $G_s = F_{s0}$, where F_{s0} is the free energy density of the superconductor without magnetic field. Far to the right of the NS interface, i.e., in the normal metal where the magnetic field is H_{cm}, the free energy density is

$$F = F_n + H_{cm}^2/8\pi \ ,$$

where the second term is simply the energy density of the magnetic field.

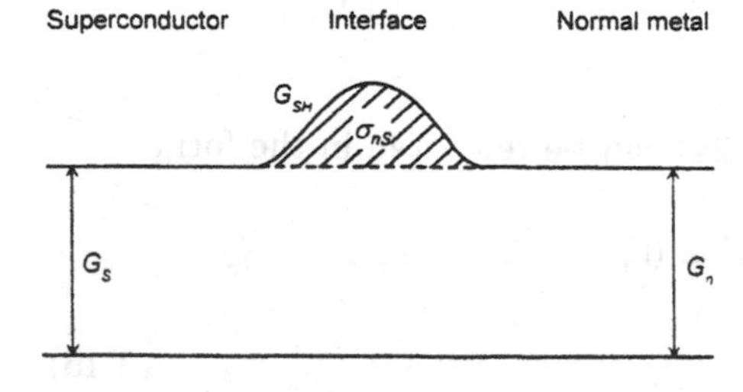

Fig. 3.3. Density of the Gibbs free energy G_{sH} in the vicinity of a normal metal–superconductor interface

In the normal domain, the Gibbs free energy density is

$$\begin{aligned} G_n &= F - HH_{cm}/4\pi = F_n + H_{cm}^2/8\pi - H_{cm}^2/4\pi \\ &= F_n - H_{cm}^2/8\pi = F_{s0} \ . \end{aligned} \quad (3.49)$$

Here we have used the well-known relation between F_n and F_{s0} (1.12):

$$F_n - F_{s0} = H_{cm}^2/8\pi$$

and the condition that in the normal domain $H = H_{cm}$.

Thus we have arrived at a result which is precisely what one should have expected: In equilibrium, the density of the Gibbs free energy far to the left of the interface equals the energy density far to the right.

But what should be happening *at* the interface? The answer is given by a sketch in Fig. 3.3.

At the interface, the Gibbs free energy density may differ from $G_{\rm n}$. Then it is natural to define the surface energy of the interface, $\sigma_{\rm ns}$, as

$$\sigma_{\rm ns} = \int_{-\infty}^{\infty} (G_{sH} - G_{\rm n})\,{\rm d}x \,, \tag{3.50}$$

where

$$G_{sH} = F_{sH} - HH_{\rm cm}/4\pi \,, \tag{3.51}$$

$$F_{sH} = F_{\rm n} + \frac{H^2}{8\pi} + \frac{H_{\rm cm}^2}{4\pi}\left[-|\psi|^2 + \frac{1}{2}|\psi|^4 + \xi^2\left|{\rm i}\nabla\psi + \frac{2\pi}{\Phi_0}A\psi\right|^2\right] , \tag{3.52}$$

$$G_{\rm n} = F_{\rm n} - H_{\rm cm}^2/8\pi \,. \tag{3.53}$$

Formula (3.51) follows from the general expression for the Gibbs free energy (where H is the magnetic field in the superconductor), formula (3.52), which forms the basis of the GL theory, can be easily derived from (3.7), and (3.53) was obtained earlier (see (3.49)).

On substitution of (3.51–53) into (3.50), we have

$$\sigma_{\rm ns} = \int_{-\infty}^{\infty}\left\{\frac{H_{\rm cm}^2}{4\pi}\left[-|\psi|^2 + \frac{1}{2}|\psi|^4 + \xi^2\left|{\rm i}\nabla\psi + \frac{2\pi}{\Phi_0}A\psi\right|^2\right] + \frac{H^2}{8\pi} - \frac{HH_{\rm cm}}{4\pi} + \frac{H_{\rm cm}^2}{8\pi}\right\}{\rm d}x \,.$$

After recalling that $\boldsymbol{A} = (0, A, 0)$ and ψ is real, this becomes

$$\sigma_{\rm ns} = \int_{-\infty}^{\infty}\left\{\frac{H_{\rm cm}^2}{4\pi}\left[-\psi^2 + \frac{1}{2}\psi^4 + \xi^2\left(\frac{{\rm d}\psi}{{\rm d}x}\right)^2 + \left(\frac{2\pi\xi A}{\Phi_0}\right)^2\psi^2\right] + \frac{H^2}{8\pi} - \frac{HH_{\rm cm}}{4\pi} + \frac{H_{\rm cm}^2}{8\pi}\right\}{\rm d}x \,.$$

Finally, using (3.48), we obtain

$$\sigma_{\rm ns} = \frac{H_{\rm cm}^2}{2\pi}\int_{-\infty}^{\infty}\left[\xi^2\left(\frac{{\rm d}\psi}{{\rm d}x}\right)^2 + \frac{H(H - H_{\rm cm})}{2H_{\rm cm}^2}\right]{\rm d}x \,. \tag{3.54}$$

Let us analyze the result. First of all, the field penetrating the superconducting region is always less than the field at the interface, that is, less than $H_{\rm cm}$. Therefore, the second term in square brackets is always negative.

Now we understand why $\sigma_{ns} < 0$ in the London theory: because quantum effects are not taken into account and, consequently, there is no term $\xi^2 (d\psi/dx)^2$. One can see from (3.54) that this inadequacy of the London theory is removed by the GL theory. By taking into account quantum effects, a positive term $\xi^2 (d\psi/dx)^2$ has appeared that leads to the energy σ_{ns} being positive.

Let us now make some evaluations. On going from N to S, the order parameter changes from 0 to 1 in the vicinity of the interface. This change takes place over a distance of the order of the coherence length ξ. Therefore, $d\psi/dx \sim 1/\xi$ and $\xi^2(d\psi/dx)^2 \sim 1$. This term is nonzero over a distance $x \sim \xi$ near the interface. Therefore,

$$\int_{-\infty}^{\infty} \xi^2 (d\psi/dx)^2 dx \sim \xi \, . \tag{3.55}$$

The term $H(H - H_{cm})/2H_{cm}^2$ reaches approximately -1 at the interface and becomes zero deep in the interior of both S and N domains. The area where it is nonzero extends over a distance of the order of the penetration depth λ. Hence the contribution of this term to the integral in (3.54) is roughly $-\lambda$.

Consider two limiting cases.

(1) $\kappa \ll 1$, i.e., $\lambda \ll \xi$. Then, by (3.55), the dominant contribution to the integral in (3.54) comes from the gradient term and

$$\sigma_{ns} \sim H_{cm}^2 \xi > 0 \, .$$

Exact integration of (3.54) using the GL theory yields

$$\sigma_{ns} = 1.89 \, \frac{H_{cm}^2}{8\pi} \, \xi \, . \tag{3.56}$$

(2) $\kappa \gg 1$, i.e., $\lambda \gg \xi$. In this case, the dominant contribution to the integral in (3.54) comes from the term $H(H - H_{cm})/2H_{cm}^2$ and the surface energy of the interface is

$$\sigma_{ns} \sim -H_{cm}^2 \, \lambda \, .$$

Exact calculations yield

$$\sigma_{ns} = -\frac{H_{cm}^2}{8\pi} \, \lambda \, . \tag{3.57}$$

Let us now interpret the results physically.

(1) The case $\kappa \ll 1$, $\lambda \ll \xi$. Figure 3.4 shows how the order parameter ψ and the magnetic field H vary in the vicinity of the interface. The former falls off over a distance ξ and the latter over a distance λ. As a result, there is a region of thickness $\sim \xi$ where the order parameter is already sufficiently small and the magnetic field is kept out. This region enjoys 'the privilege' of the superconductor: it is free of magnetic field; but the order parameter is still very small there compared to the superconducting regions. This ought to result in an increase of the region's energy compared to the superconducting

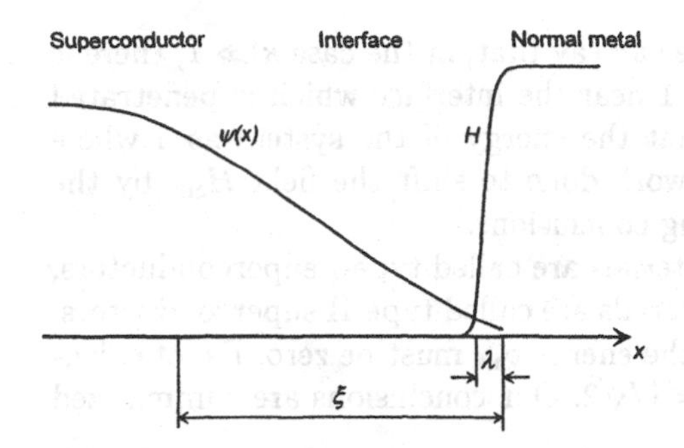

Fig. 3.4. Spatial variations of the order parameter ψ and the magnetic field H in the vicinity of the NS interface for $\kappa \ll 1$

area at the far left. In other words, the energy of this region exceeds that of the superconducting region by the additional energy required to break the electron (Cooper) pairs within the region and thus to reduce the order parameter ψ. The density of the additional energy is $H_{\text{cm}}^2/8\pi$, hence the energy of the region is $\sim H_{\text{cm}}^2\xi/8\pi$, in accord with (3.56).

This situation can be looked at from a different perspective. To keep the magnetic field out of the region with a reduced value of ψ (i.e., with the energy close to the energy of the normal metal), work must be done on the magnetic field to expel it from this region. This means that one must overcome a 'pressure' from the magnetic field $H_{\text{cm}}^2/8\pi$ and shift its boundary by the distance ξ to the right. For this, the work $(H_{\text{cm}}^2/8\pi)\xi$ must be done.

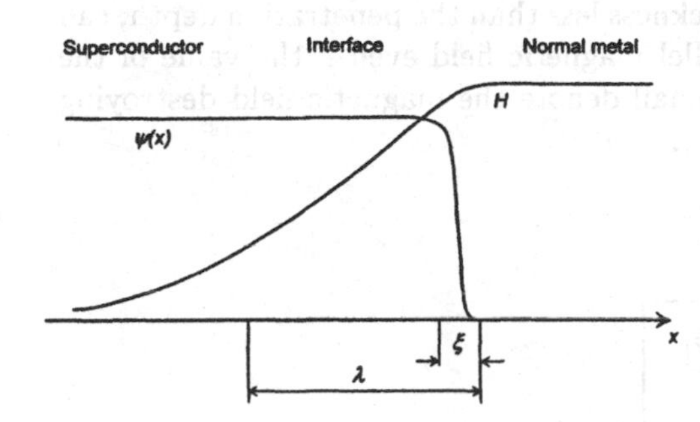

Fig. 3.5. Spatial variations of the order parameter ψ and the magnetic field H in the vicinity of the NS interface for $\kappa \gg 1$

(2) The case $\kappa \gg 1$, $\lambda \gg \xi$. The variations of $\psi(x)$ and $H(x)$ in this case are sketched in Fig. 3.5. This time, ψ varies much more rapidly than the magnetic field so that there is a region of thickness $\sim \lambda$ where the order parameter is $\psi \sim 1$ but a relatively strong magnetic field still remains. The presence of the magnetic field compels us to compare this region with the normal metal. Unlike the latter, the electrons in the region are coupled in Cooper pairs ($\psi \sim 1$). Hence its energy is less than the energy of the normal region to the right by the value of the condensation energy. As the thickness of the region is $\sim \lambda$ and the condensation energy density is $H_{\text{cm}}^2/8\pi$, it is evident that $\sigma_{\text{ns}} \sim -(H_{\text{cm}}^2/8\pi)\lambda$.

From a different perspective, one can say that, in the case $\kappa \gg 1$, there is a region of thickness $\sim \lambda$ with $\psi \sim 1$ near the interface which is penetrated by the magnetic field. It implies that the energy of the system as a whole has decreased by the value of the work done to shift the field H_{cm} by the distance λ. We come to the following conclusions.

If $\kappa \ll 1$, then $\sigma_{ns} > 0$. Such materials are called type-I superconductors.

If $\kappa \gg 1$, then $\sigma_{ns} < 0$. Such materials are called type-II superconductors.

Evidently, at some value $\kappa \sim 1$, the energy σ_{ns} must be zero. Exact calculations show that this occurs at $\kappa = 1/\sqrt{2}$. Our conclusions are summarized in Table 3.1:

Table 3.1.

Type I	Type II
$\kappa < 1/\sqrt{2}$	$\kappa > 1/\sqrt{2}$
$\sigma_{ns} > 0$	$\sigma_{ns} < 0$

3.5 Critical Field of a Thin Film

A thin superconducting film, of thickness less than the penetration depth, can remain superconducting in a parallel magnetic field even if the value of the field is well in excess of H_{cm}. We shall denote the magnetic field destroying the film's superconductivity by $H_{c\parallel}$.

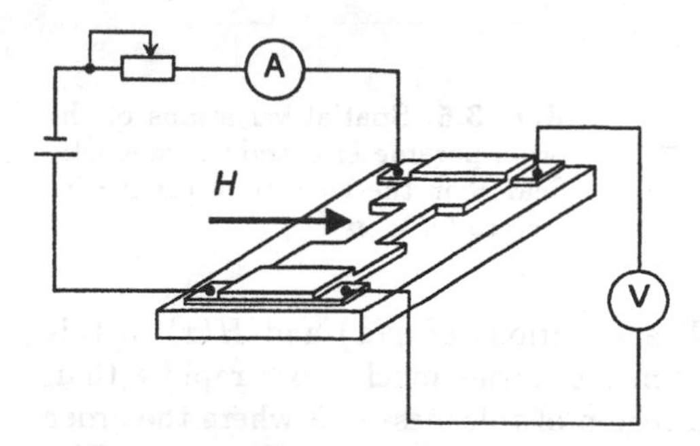

Fig. 3.6. Scheme of the critical magnetic field measurements on a thin film

Consider an experiment shown schematically in Fig. 3.6. A thin film of the material under study is deposited onto a clean glass substrate and connected to a measuring circuit by current and voltage leads. Then the specimen is mounted in a cryostat in such a way that an external magnetic field is parallel to its surface. The external field is increased and at a certain value of the

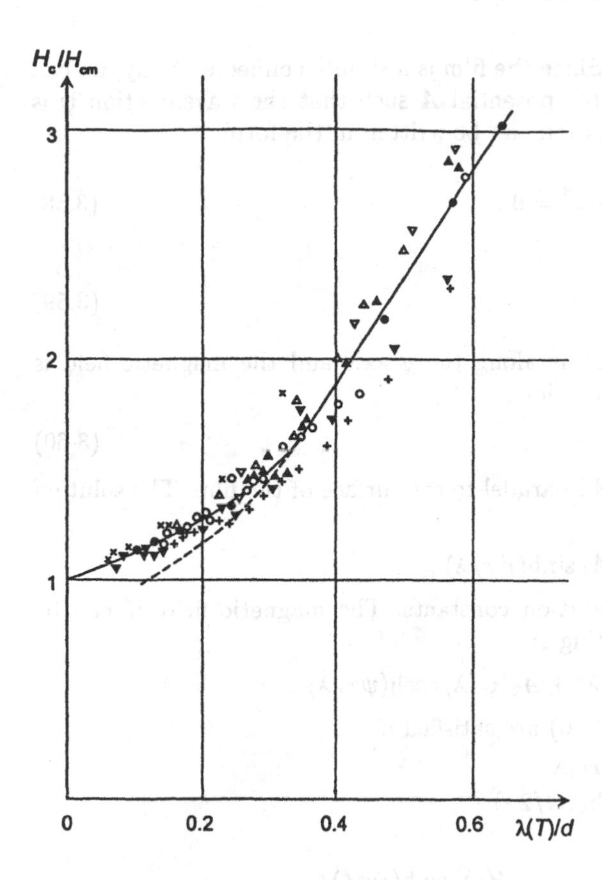

Fig. 3.7. Results of the critical magnetic field measurements on Sn films of various thicknesses [29]; d is the film thickness. The solid curve shows the result of the Ginzburg–Landau theory

field the film begins to show a finite resistance. This is the critical field of the film. A compilation of the results of such an experiment is shown in Fig. 3.7.

We remember from Sect. 1.3 that a bulk superconductor in an external field H_{cm} undergoes a first-order phase transition. Indeed, one notices in Fig. 3.7 that thick films show supercooling. This means that, when the film is in the normal state and the field decreases, the transition to the superconducting state may not occur at the appropriate value of the field, that is, at the field for which the free energies of the normal and superconducting states are equal. The transition may be delayed down to a lower field called the supercooling field. In Fig. 3.7 the supercooling fields are shown by the dashed curve. Thus we observe that, as the film thickness decreases, the effect of supercooling becomes weaker and weaker and finally, starting from a certain thickness, disappears completely. This leads to a natural conclusion: that the superconducting transition for films of this thickness or thinner is of second order.

Assume now that the thickness of our film is $d \ll \xi, \lambda$. Then we may neglect variations of ψ in the film and assume that the magnetic field penetrates the film almost completely. The surfaces of the film are assumed to coincide with the planes $x = \pm d/2$. Since the film is a simply connected body, we can choose a gauge for the vector potential $\boldsymbol{A}$ such that the wavefunction ψ is real. Then the first GL equation can be written in the form

$$-\left[1 - \left(\frac{2\pi\xi A}{\Phi_0}\right)^2\right]\psi + \psi^3 = 0\,, \tag{3.58}$$

$$\frac{\mathrm{d}^2 A}{\mathrm{d}x^2} = \frac{\psi^2}{\lambda^2} A\,. \tag{3.59}$$

Here the vector potential $\boldsymbol{A}$ is along the y axis and the magnetic field is subject to the boundary condition

$$H(\pm d/2) = H_0\,, \tag{3.60}$$

where H_0 is the external field parallel to the surface of the film. The solution of (3.59) is

$$A = A_1 \cosh(\psi x/\lambda) + A_2 \sinh(\psi x/\lambda)\,,$$

where A_1 and A_2 are integration constants. The magnetic field H can be found simply by differentiating A:

$$H = A_1(\psi/\lambda)\sinh(\psi x/\lambda) + A_2(\psi/\lambda)\cosh(\psi x/\lambda)\,.$$

The boundary conditions (3.60) are satisfied if

$$A_1 = 0\,, \quad A_2 = \frac{H_0\lambda}{\psi\cosh(\psi d/2\lambda)}\,.$$

Then

$$H = H_0\frac{\cosh(\psi x/\lambda)}{\cosh(\psi d/2\lambda)}\,, \qquad A = \frac{H_0\lambda}{\psi}\frac{\sinh(\psi x/\lambda)}{\cosh(\psi d/2\lambda)}\,. \tag{3.61}$$

Let us now apply the results to a thin film ($d \ll \lambda$). In this case, $\psi x/\lambda \ll 1$ and $\psi d/2\lambda \ll 1$ and we can expand the hyperbolic functions in a Taylor's series. Keeping only linear terms, we have

$$\cosh(\psi d/2\lambda) = 1\,, \qquad \sinh(\psi x/\lambda) = \psi x/\lambda\,.$$

On substituting this into (3.61), we obtain: $A = H_0 x$ and then, from (3.58),

$$\psi^2 = 1 - \left(\frac{2\pi\xi}{\Phi_0}\right)^2 {H_0}^2 x^2\,.$$

To find the average of the last expression over the thickness of the film, we integrate with respect to x from $-d/2$ to $+d/2$:

$$\psi^2 d = d - {H_0}^2 \frac{1}{12} d^3 \left(\frac{2\pi\xi}{\Phi_0}\right)^2\,.$$

Taking into account (3.38), we finally obtain the dependence of the order parameter in the film on the applied magnetic field:

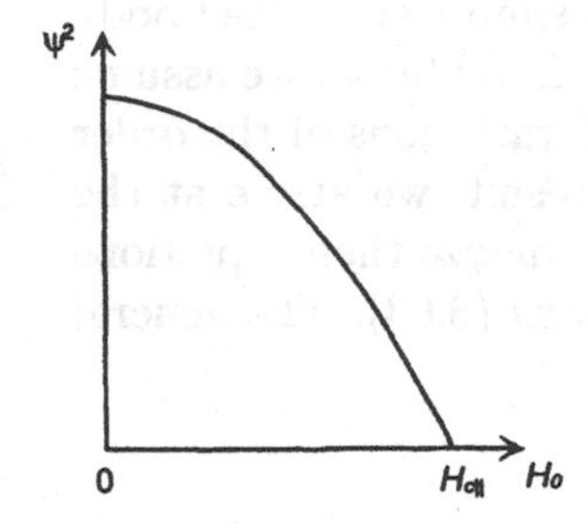

Fig. 3.8. Order parameter as a function of the parallel external magnetic field H_0 for a thin film ($d \ll \lambda$)

$$\psi^2 = 1 - \frac{1}{24}\frac{{H_0}^2 d^2}{H_{\text{cm}}^2 \lambda^2} .$$

This dependence is sketched in Fig. 3.8. Thus, the order parameter ψ in a thin film is strongly dependent on the applied field H_0 and gradually goes to zero at the value of the field of $2\sqrt{6}H_{\text{cm}}\lambda/d$. It is natural to select this field as the critical field of the film, $H_{\text{c}\parallel}$.

Finally, the critical field of a thin film of thickness d in a parallel external field H_0 is

$$H_{\text{c}\parallel} = 2\sqrt{6}\,H_{\text{cm}}\,\frac{\lambda}{d} . \tag{3.62}$$

Thus, as the thickness of the film decreases, its critical field goes up. For example, if the thickness of a film is an order of magnitude less than the penetration depth, $\lambda/d \sim 10$, and $H_{\text{cm}} \sim 10^3$ Oe, we obtain, by (3.62), $H_{\text{c}\parallel} \sim$ 40 000 Oe. Physically, this is easy to understand. Since the magnetic field penetrates into the film, the film's diamagnetic moment per unit volume is substantially reduced compared to a bulk specimen. But a small diamagnetic moment in an external field is equivalent to a small magnetic needle oriented against the field. Such a situation is energetically more favorable than in the case of a large diamagnetic moment. Therefore, a film in an external field is much more stable than a bulk specimen and can remain superconducting up to much higher fields.

3.6 Critical Current of a Thin Film

Let us now consider a film carrying a current I, in the absence of external magnetic field. We assume, as before, that the surfaces of the film coincide with the planes $x = \pm d/2$ and that the current flows in the y direction. The current I is the total current through the film per unit length of the z axis.

The current generates a magnetic field H_I at the surfaces of the film. This imposes the boundary conditions

$$H(\pm d/2) = \mp H_I . \tag{3.63}$$

As in the preceding section, we note that the film is a simply connected body and use this to choose a gauge for $\boldsymbol{A}$ so that ψ is real. In addition, we assume that the film is thin: $d \ll \lambda$, ξ. Then we may neglect variations of the order parameter over the thickness of the film and, as a result, we arrive at the GL equations of the form (3.58) and (3.59). We shall analyze these equations again but this time subject to the boundary conditions (3.63). The general solution of (3.59) is

$$A = A_1 \cosh(\psi x/\lambda) + A_2 \sinh(\psi x/\lambda) .$$

Since for our problem $H = \mathrm{d}A/\mathrm{d}x$, we have

$$H = (A_1\psi/\lambda) \sinh(\psi x/\lambda) + (A_2\psi/\lambda) \cosh(\psi x/\lambda) .$$

On substitution of the boundary conditions (3.63) into this expression we obtain two equations with two unknowns (A_1 and A_2). Their solution is

$$A_1 = \frac{\lambda H_I}{\psi \sinh(\psi d/2\lambda)} , \qquad A_2 = 0 .$$

The final result is

$$H = H_I \frac{\sinh(\psi x/\lambda)}{\sinh(\psi d/2\lambda)} , \qquad A = \frac{\lambda H_I \cosh(\psi x/\lambda)}{\psi \sinh(\psi d/2\lambda)} . \tag{3.64}$$

Now we take into account that the thickness of the film is small, i.e., $\psi d \ll \lambda$. This implies that in the expression for A in (3.64) we can assume $\cosh(\psi x/\lambda) \approx 1$ and $\sinh(\psi d/2\lambda) \approx \psi d/2\lambda$. Then

$$A = 2\lambda^2 H_I/\psi^2 d .$$

Substituting this into (3.58) and using (3.38), we obtain

$$\frac{2\lambda^2 {H_I}^2}{d^2 H_{\mathrm{cm}}^2} = \psi^4 - \psi^6 . \tag{3.65}$$

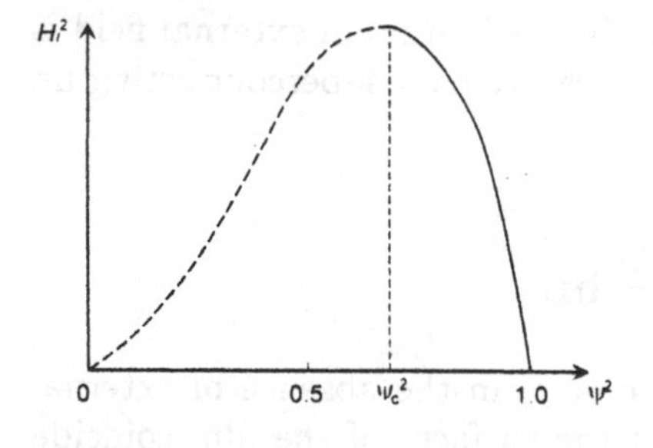

Fig. 3.9. Order parameter in a thin superconducting film as a function of the applied current. The heavy part of the curve corresponds to stable superconductivity

The dependence of ${H_I}^2$ on ψ^2 is plotted in Fig. 3.9. The heavy line emphasizes the part of the plot corresponding to stable states. When the film does not carry any current, equation (3.65) has two solutions: $\psi = 0$ and $\psi = 1$. From general arguments, it is clear that in this case the superconducting state ($\psi = 1$) is more favorable energetically and, therefore, it is this that will be realized rather than the normal state. If we now switch on a very small current across the film, equation (3.65) will again have two solutions: one corresponding to $\psi \le 1$, and the other corresponding to $\psi \ll 1$. It then follows from the requirement of continuity that it is the former that corresponds to the stable superconducting state.

But what is the maximum current that the film can sustain while remaining in the superconducting state?

The answer follows directly from the plot of Fig. 3.9: it is the current corresponding to the maximum in the plot. Let us work out its value. First, we find ψ_c corresponding to the maximum of $H_I{}^2$:

$$\frac{\mathrm{d}}{\mathrm{d}\psi}(2\lambda^2 H_I{}^2/d^2 H_{\mathrm{cm}}^2) = 4\psi_c{}^3 - 6\psi_c{}^5 = 0 ,$$

which leads to $\psi_c{}^2 = 2/3$.
Then

$$H_{I_c} = \frac{\sqrt{2}}{3\sqrt{3}} H_{\mathrm{cm}} \frac{d}{\lambda} . \tag{3.66}$$

From (3.66) and the formula $H_{I_c} = (4\pi/c)\, I_c$, one can easily find the average critical current density j_c across the film:

$$j_c = \frac{\sqrt{2}}{12\pi\sqrt{3}} \frac{cH_{\mathrm{cm}}}{\lambda} . \tag{3.67}$$

Let us summarize the major results of this section.

(1) The field generated by the critical current at the surface of a film is proportional to the thickness of the film: $H_{I_c} \propto d$, namely, it becomes smaller when the thickness of the film decreases. At the same time, the critical field of the film varies as $H_{c\parallel} \propto 1/d$, that is, it increases when the film becomes thinner. For example, if $d/\lambda \sim 0.1$ and $H_{\mathrm{cm}} = 1000$ Oe, then $H_{c\parallel} \sim 4\times10^4$ Oe and $H_{I_c} \sim 30$ Oe. Hence, for a thin film, the destruction of superconductivity by a current is not identical to its destruction by the magnetic field of this current.

(2) The critical current density j_c from (3.67) is independent of the thickness of the film. It follows that j_c is simply a characteristic of the current-carrying capacity of a particular material.

(3) Finally, let us emphasize one more important point. The destruction of superconductivity by a current is not accompanied by a phase transition. This is only true, of course, provided that the film is in good thermal contact with liquid helium such that all the heat given up by the film is immediately thermalized. Then the temperature of the film remains, at any moment, equal to the temperature of the helium bath. In this case, even if the current reaches the value of I_c, the free energy of the superconducting state remains below the free energy of the normal state.

What happens then when the current is $I = I_c$? Why cannot a supercritical current pass through the film without dissipation? To answer these questions, we shall use the analysis proposed by Bardeen. Since the film is thin, we can neglect the energy of the magnetic field generated by the current and write the free energy density in the form

$$F_s = F_n - |\alpha|\, n_s + \frac{\beta}{2} {n_s}^2 + n_s \frac{m}{2} {v_s}^2 .$$

Here n_s is the superconducting electron density; α and β are the coefficients of the GL theory, which are already known to us. The last term describes the kinetic energy of the superconducting electrons and v_s is the velocity of their correlated motion, or 'superfluid' velocity.

The equilibrium value of $n_s(v_s)$ as a function of v_s can be found from the requirement that the free energy at equilibrium is a minimum:

$$\frac{\partial F_s}{\partial n_s} = -|\alpha| + \beta n_s + \frac{m {v_s}^2}{2} = 0 .$$

This yields

$$n_s = \frac{\left(|\alpha| - \dfrac{m {v_s}^2}{2}\right)}{\beta} . \tag{3.68}$$

The current density j_s in this case is determined by the well-known expression

$$j_s = n_s e v_s . \tag{3.69}$$

The dependences $n_s(v_s)$ and $j_s(v_s)$, corresponding to (3.68) and (3.69), respectively, are sketched in Fig. 3.10. Again, the part of the curve $j_s(v_s)$ corresponding to stable states is shown in bold. What does this plot tell us? As the superconducting current density increases, the electron velocity also increases. But this is accompanied by a decrease of the electron density $n_s(v_s)$. The electron pairs are being broken. Finally, a state is attained such that it is not possible to increase the superconducting current any further, because the superconducting electron density becomes so low. There are simply not enough carriers to maintain a higher current. This is the critical current given by (3.67). It is often called the pair-breaking current.

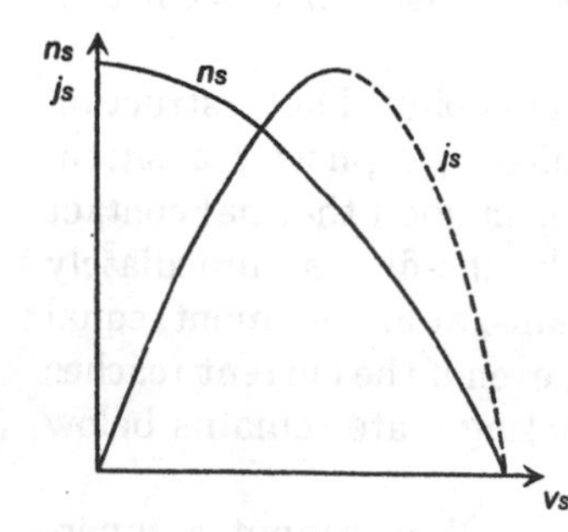

Fig. 3.10. Superconducting electron density n_s and supercurrent density j_s in a thin film as functions of the velocity of the superconducting component of the electron density

Problems

Problem 3.1. Consider a Pb cylinder at 4.2 K in a uniform magnetic field parallel to its axis. The field at the surface of the cylinder is $H_0 = 300$ Oe. Find the density w of the magnetic energy in the specimen at the distance $x = 300$ Å from the surface. (The diameter of the cylinder is much larger than the magnetic field penetration depth.)

Problem 3.2. The critical temperature of Pb is 7.18 K. Find the ratio of the value of the penetration depth λ at $T = 7.10$ K to its value at $T = 4.2$ K. Estimate the density of the superconducting electrons at $T = 7.10$ K.

Problem 3.3. Consider a specimen of superconducting Sn. At $T = 0.9\,T_c$, $\lambda = 8.7 \times 10^{-6}$ cm and $\xi = 4.35 \times 10^{-5}$ cm. Find the surface energy of the interface between a normal and a superconducting region, σ_{ns}.

Problem 3.4. Consider a Sn film of thickness $d = 1000$ Å deposited onto a glass substrate. The film is placed in an external magnetic field $H_0 = 10$ Oe parallel to its surface. The temperature is $T = 0.9\,T_c$. Find the field at the center of the film and the diamagnetic moment M_0 per unit area of the film's surface.

Problem 3.5. Consider a thick Sn film in a parallel magnetic field. The thickness of the film is 10^{-4} cm and the temperature is $T = 0.9\,T_c$. Find the critical field of the film, $H_{c\|}$, on the assumption that the order parameter ψ is independent of the field and equals 1 (the first-order phase transition).

Problem 3.6. Consider a Pb film of thickness $d = 200$ Å deposited onto a glass substrate. At $T = 0$, the magnetic field penetration depth is $\lambda(0) = 390$ Å, and the thermodynamic critical field is $H_{cm}(0) = 803$ Oe. Find the magnetic field generated by the critical current at the surface of the film. Compare it with the parallel critical field of the same film.

Problem 3.7. Find the critical current density for the film of the previous problem.

Problem 3.8. Consider a thin film ($d \ll \lambda$) of In in the superconducting state. The GL parameter of the film is $\kappa = 0.1$. The magnetic field penetration depth is$\lambda = 800$ Å. Find the critical velocity of the superconducting electrons.

Problem 3.9. Consider a bulk type-I superconductor in an external magnetic field equal to the thermodynamic critical field H_{cm}. Compare the density of the Meissner current j_M at the surface of the superconductor with the critical current density j_c of a thin film made of the same material.

4. Weak Superconductivity

4.1 Phase Coherence and Types of Weak Links

In 1962 B. Josephson published a theoretical paper [10] predicting the existence of two fascinating effects. One was supposed to find them in superconducting tunnel junctions. The basic idea of the first effect was that a tunnel junction[1] should be able to sustain a zero-voltage (superconducting) current. The critical value of this current was predicted to depend on the external magnetic field in a very unusual way: If the current exceeds its critical value, which is a characteristic of a particular junction, the junction begins to generate high-frequency electromagnetic waves. This is the second Josephson effect.

Both effects were thoroughly verified by experiment [11, 30] shortly after [10] was published. Moreover, it soon became clear that the Josephson effects exist not only in tunnel junctions, but also in other kinds of the so-called weak links, that is, short sections of superconducting circuits where the critical current is substantially suppressed [20, 31, 32].

The effects of weak superconductivity have their origin in the quantum nature of the superconducting state. We already know that the foundation of the superconducting state is the existence of the Bose condensate. That is to say, all electron pairs in the superconducting state occupy the same quantum level and are described by a single wavefunction, common to all of them. Their behavior is mutually conditioned, they are coherent.

Imagine two bulk pieces of a superconductor at the same temperature, completely isolated from each other. When both pieces are in the superconducting state, the superconducting electrons in each of them have their own superconducting wavefunction. Furthermore, since the temperatures and the materials of the superconductors are identical, the amplitudes of the wavefunctions must also be the same. This does not hold for the phases, however, which are arbitrary. This situation remains as long as the superconductors are isolated from each other. Let us now establish a weak contact between them, i.e., a contact that is weak enough so as not to change radically the electron states of the two pieces but sufficient to play the role of a perturba-

[1] Concerning electron tunneling between two superconductors separated by a thin insulating barrier, see Sect. 6.4.

tion. A new wavefunction will then emerge for the superconductor as a whole, which can be considered as a result of interference between the wavefunctions from the two pieces. As mentioned above, the amplitudes of the two wavefunctions had been equal even before the weak link was established. But not their phases: the phase coherence is a direct result of establishing the weak link. It is therefore often said that phase coherence sets up in weakly-coupled superconductors.

Here it should be noted that a weak link between two superconductors is merely a convenient place for letting interference effects become visible. As to the interference effects in superconductors themselves, they were known long before the discovery of the Josephson effects. The most famous example is the magnetic flux quantization in a superconducting ring: the magnetic flux frozen in such a ring can assume only quantized values because the supercurrent, circulating in the ring and generating this magnetic flux, is quantized. And the quantization of the supercurrent is a typical interference effect. The current can assume only those values that yield an integral number of wavelengths of the superconducting wavefunction over the length of the ring. This situation is exactly analogous to the quantization of electron orbits in the Bohr atom.

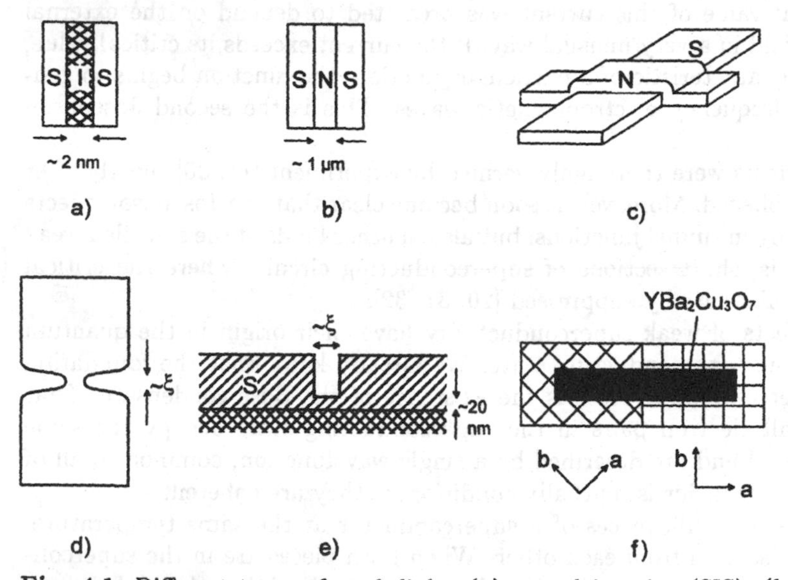

Fig. 4.1. Different types of weak links: **(a)** tunnel junction (SIS); **(b)** sandwich (SNS); **(c)** normal film N causes local suppression of the order parameter of a superconducting film S; **(d)** Dayem bridge, top view; **(e)** bridge of variable thickness, longitudinal cross-section; **(f)** grain-boundary junction

Let us now consider different types of weak links. First, there are devices without concentration of current such as tunnel junctions (see Fig. 4.1 (a)).

The thickness of an insulating layer is typically about 1–2 nm and the critical current density is in the range 10–10^4 A cm^{-2}, i.e., much less than the critical current density of the bulk superconductors.[2]

In superconductor–normal metal–superconductor (SNS) sandwiches, the normal layer can be as thick as 10^{-4} cm (Fig. 4.1 (b)). The wavefunctions of the superconducting electrons penetrate the normal metal due to the proximity effect. In the region of their overlap, the wavefunctions interfere, with the consequence that phase coherence is established between the bulk superconductors. If the amplitude of the superconducting wavefunction in the weak link is small, the critical current is also small.

The same effect can be achieved if the normal layer between the two superconductors is replaced by a doped semiconductor or another superconductor with a small critical current density. For example, if a narrow superconducting film is covered with a narrow film of a normal metal (Fig. 4.1 (c)), the amplitude of the superconducting electron wavefunction in the film is reduced where the film is in contact with the normal metal, due to the proximity effect. This causes a local decrease of the critical current density, that is, a weak link is established.

In devices with concentration of current, the critical current density in the weak link is the same as in the bulk, but the absolute value of the critical current is much less. A superconducting film with a short narrow constriction (Dayem bridge) falls into this category, provided the size of the constriction is of the order of the coherence length ξ (see Fig. 4.1 (d)). Another example is a bridge of variable thickness, such that the thickness of the main film is hundreds of nm while the thickness of the bridge itself is only several dozen nm (Fig. 4.1 (e)).

Finally, a weak link typical of high-temperature superconductors is shown in Fig. 4.1 (f). It is called a grain boundary (or bi-crystal) junction. Due to the extremely short coherence length in high-T_c materials ($\xi \sim 1$ nm), defects in their crystal structure can act as weak links. The best-controlled defects can be produced between two regions of an epitaxially grown high-T_c film with different crystal orientations (grain boundary). The critical current density of such a weak link can be varied by changing the misorientation angle between the two crystallites.

Different types of weak links based on conventional low-T_c superconductors are described in a classic review by Likharev [31]. Various weak links and SNS structures made of high-T_c superconducting materials have been recently reviewed by Gross [33].

[2] High-quality tunnel Josephson junctions, typically made of niobium with a barrier of aluminum oxide, presently serve as basic elements for low-temperature superconducting electronics. Their critical current densities can be as high as 10^4–10^5 A cm^{-2}.

4.2 The DC Josephson Effect

Let us start with the first Josephson effect which is a dc effect. The physics behind it is essentially as follows. A sufficiently small current can pass through a weak link (Josephson junction) without dissipation. In other words, when such a current passes through the weak link, no voltage is generated across the junction. Since the current is always small, the magnetic field generated by it can be neglected and, following the Ginzburg–Landau theory, we can argue that the density of the current is determined by the phase gradient $\nabla\theta$ of the superconducting electron wavefunction.

An important characteristic of the weak link is that the phase gradient $\nabla\theta$ is very large compared to the phase gradients inside the bulk superconductors (in the following we shall refer to them as 'electrodes'). For a tunnel junction, strictly speaking, the expression 'phase gradient' is inappropriate and one should rather speak of a discontinuous phase jump across the junction. Therefore, from now on we shall always analyze the *phase difference* across the weak link:

$$\varphi = \theta_2 - \theta_1 ,$$

where θ_1 and θ_2 are the phases of the superconducting electron wavefunctions in the first and second electrodes, respectively.

Now our task is to find the relation between the current through the weak link, I_s, and the phase difference φ. Let us first establish some general and almost self-evident relations.

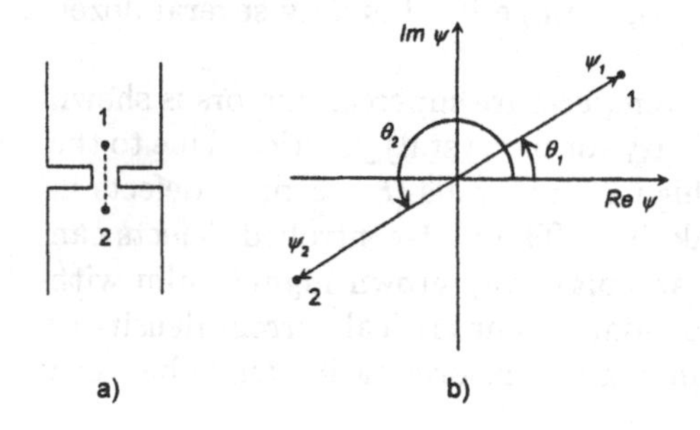

Fig. 4.2. **(a)** Example of weak link (Dayem bridge). The phase gradient $\nabla\theta$ is taken along the dashed line from point *1* to point *2*; **(b)** representation of the order parameter in the complex plane. Passing from point *1* to point *2* in the weak link corresponds to passing from ψ_1 to ψ_2 in the complex plane, along the line *1*–*2*

(1) If the current through the junction is $I_s = 0$, the phase difference is also $\varphi = 0$.

(2) Since the variation of the phase θ by 2π in one of the electrodes does not change anything physically, it is evident that $I_s(\varphi)$ is a periodic function with period 2π, that is, $I_s(\varphi) = I_s(\varphi + 2\pi)$.

(3) Changing the sign of the current must cause a change of the sign of the phase difference; therefore, $I_s(\varphi) = -I_s(-\varphi)$.

(4) The last relation, $I_s(\pi) = 0$, is somewhat less obvious. Let us assume[3] that the length of the weak link is small but finite so that we can introduce a phase gradient $\nabla\theta$ across it. On going along the dashed line (Fig. 4.2 (a)) from point *1* to point *2*, the complex order parameter runs through values varying from ψ_1 to ψ_2 along the straight line *1–2* in the complex plane of ψ (see Fig. 4.2 (b)). It is obvious that $\nabla\theta$ along the line is zero, resulting in $I_s = 0$. At this point one can raise an objection. What happens if we go from ψ_1 to ψ_2 not along the straight line but along, say, a semicircle of radius $|\psi|$? The answer is $\nabla\theta \neq 0$ and $I_s \neq 0$. Such a situation indeed occurs in long thin superconducting filaments. In that case, however, two paths are possible to go from ψ_1 to ψ_2: anticlockwise and clockwise, so that the corresponding gradients $\nabla\theta$ will be of opposite signs. This means that $I_s(\varphi)$ will become a multiple-valued function. However, by definition, such junctions can no longer be regarded as Josephson junctions.

Now let us ask: what is the simplest representation of $I_s(\varphi)$ having all four of the above properties? The function $\sin\varphi$ is an obvious satisfactory answer, and we can propose

$$I_s(\varphi) = I_c \sin\varphi\,, \tag{4.1}$$

where I_c is the maximum dissipation-free current through the junction often referred to as the critical current.

Let us prove that our proposal is indeed appropriate. Two derivations of (4.1) will be given. The first, applicable to tunnel Josephson junctions, is due to Feynman [34] and the second, relevant to the case of short narrow constrictions, is by Aslamazov and Larkin [35].

But first a short foray into quantum mechanics.

The evolution of a quantum-mechanical system with time is described by a wavefunction $\Psi(t)$ which is the solution of the Schrödinger equation:

$$i\hbar\frac{\partial\Psi}{\partial t} = \hat{H}\Psi\,, \tag{4.2}$$

where $\hat{H}$ is the Hamilton operator of the system. If the system is allowed only discrete states ψ_α (where α is a series of indices each characterizing a given state), its wavefunction can be expanded in a series

$$\Psi(t) = \sum_\alpha C_\alpha(t)\,\psi_\alpha\,. \tag{4.3}$$

Substitution of this in (4.2) gives

$$i\hbar\frac{\mathrm{d}C_\beta}{\mathrm{d}t} = \sum_\alpha H_{\beta\alpha}C_\alpha(t)\,. \tag{4.4}$$

Here

$$H_{\beta\alpha} = \int \psi_\beta^*\hat{H}\psi_\alpha\,\mathrm{d}V\,.$$

[3] The following argumentation is due to K.K. Likharev.

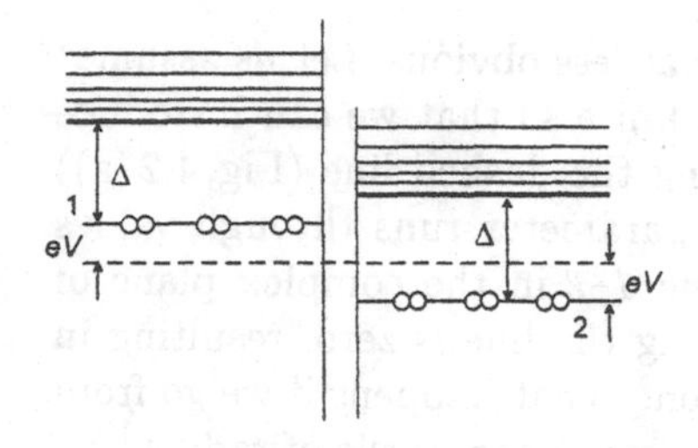

Fig. 4.3. Energy-level diagram for a tunnel Josephson junction with a finite voltage V applied to it. Levels *1* and *2* are separated by $2eV$

One can see that $H_{\beta\beta}$ is the energy of the system in the state ψ_β, and $H_{\beta\alpha}$ is the matrix element characterizing the probability of a transition from the state ψ_α to the state ψ_β. The function $C_\alpha(t)$ represents the amplitude of the state ψ_α, and $|C_\alpha|^2$ gives the probability of finding the system in the state ψ_α.

Let us now go back to a tunnel Josephson junction. Assume that, in the general case, the current through the junction is so large ($I_s > I_c$) that a voltage V appears across the junction. The corresponding energy diagram is shown in Fig. 4.3. Following Feynman [34], we consider a system of superconducting electrons, or Cooper pairs, as a two-level quantum-mechanical system. That is, we suppose that an electron pair with charge $2e$ can occupy either level *1* or level *2*. Then its energy will be either H_{11} or H_{22}, respectively, where $H_{11} = eV$ and $H_{22} = -eV$. The transition from level *1* to level *2* is governed by the matrix element $H_{12} = H_{21} = K$. Then (4.4) becomes

$$\begin{aligned} i\hbar \frac{dC_1}{dt} &= eVC_1(t) + KC_2(t) \,, \\ i\hbar \frac{dC_2}{dt} &= KC_1(t) - eVC_2(t) \,. \end{aligned} \tag{4.5}$$

Here C_1 is the amplitude of the pair state at level *1* and $|C_1|^2$ is normalized in such a way that $|C_1|^2 = n_s$, where n_s is the superconducting electron density in the junction electrodes. For the sake of simplicity, we assume that both electrodes are made of the same material. Expressing the amplitudes C_1 and C_2 as

$$C_1 = \sqrt{n_s}\, e^{i\theta_1} \,, \qquad C_2 = \sqrt{n_s}\, e^{i\theta_2} \,,$$

substituting these in (4.5), and separating real and imaginary parts, we obtain

$$\frac{dn_s}{dt} = \frac{2Kn_s}{\hbar} \sin\varphi \,, \tag{4.6}$$

$$\frac{d\theta_1}{dt} = -\frac{K}{\hbar} \cos\varphi - \frac{eV}{\hbar} \,, \tag{4.7}$$

$$\frac{d\theta_2}{dt} = -\frac{K}{\hbar} \cos\varphi + \frac{eV}{\hbar} \,, \tag{4.8}$$

where $\varphi = \theta_2 - \theta_1$.

The current through the tunnel junction is proportional to $\mathrm{d}n_s/\mathrm{d}t$. Indeed, when the current is switched on, the superconducting electron density starts to vary at the rate $\mathrm{d}n_s/\mathrm{d}t$ thereby giving rise to a current $I_s = \mathrm{d}n_s/\mathrm{d}t$. This means that electrons start to leave the superconducting electrode. But this process is immediately compensated by the arrival of new electrons from an external current source, because the junction is a part of a closed electric circuit. Therefore, the density n_s remains constant due to electroneutrality of the system as a whole. However, for defining the supercurrent, it is sufficient to assume $I_s \propto \mathrm{d}n_s/\mathrm{d}t$. Then from (4.6) we immediately obtain the equation for the dc Josephson effect

$$I_s = I_c \sin\varphi \,. \tag{4.9}$$

Another way to derive the same equation, as applied to a weak link formed by a short narrow constriction, has been proposed by Aslamazov and Larkin [35]. Assume that a bridge, as in Fig. 4.4, is so short that its length is $L \ll \xi$.

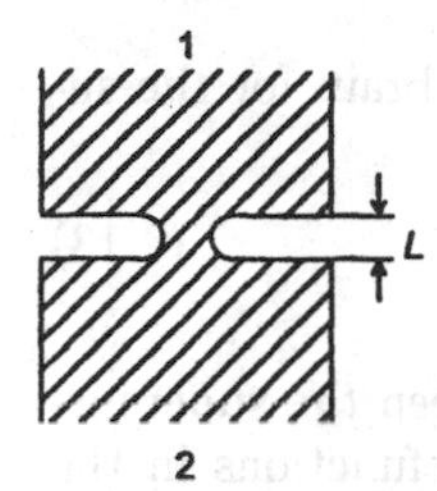

Fig. 4.4. Short narrow constriction (bridge) of length $L \ll \xi$ links together two wide films *1* and *2*

In the absence of external magnetic field, the first Ginzburg–Landau equation (3.12) for such a bridge takes the form

$$-\xi^2\nabla^2\psi - \psi + \psi\,|\psi|^2 = 0 \,. \tag{4.10}$$

The order parameter in a short bridge varies substantially over the length of the bridge, L. In this case, $\nabla^2\psi$ can be estimated as

$$\nabla^2\psi \sim \psi/L^2 \,.$$

On the other hand, the amplitude of the order parameter is $|\psi| \sim 1$. It follows that the dominant term in (4.10) is the first one, since

$$\xi^2/L^2 \gg 1 \,,$$

while all other terms are of the order 1. This allows us to simplify (4.10):

$$\nabla^2\psi = 0 \,. \tag{4.11}$$

Now assume that, far away from the constriction, the order parameter ψ in film *1* is

$$\psi = \psi_1\,\mathrm{e}^{\mathrm{i}\theta_1}$$

and in film *2* it is

$$\psi = \psi_2\,\mathrm{e}^{\mathrm{i}\theta_2} \,.$$

Here ψ_1, ψ_2, θ_1, and θ_2 are constants independent of the coordinates. Within the area of the constriction, the two wavefunctions interfere and we therefore seek the solution of (4.11) in the form

$$\psi = \psi_1 \,\mathrm{e}^{\mathrm{i}\theta_1} f(r) + \psi_2 \,\mathrm{e}^{\mathrm{i}\theta_2} [1 - f(r)] \tag{4.12}$$

subject to the conditions $f(r) \to 1$ in the interior of film *1* and $f(r) \to 0$ in the interior of film *2*.

Substituting (4.12) into (4.11), we obtain the equation for $f(r)$:

$$\nabla^2 f(r) = 0 \,.$$

However, there is no need to solve this equation here. It suffices to know that the solution exists, and we can immediately go over to calculating the supercurrent which, by the second Ginzburg–Landau equation (3.17), is

$$j_s = \frac{|\alpha|\hbar e}{\beta m} \,\mathrm{Im}(\psi^* \nabla \psi) \,.$$

Substituting here (4.12) and applying simple algebra, we obtain for the supercurrent density through the bridge:

$$j_s = j_c \sin\varphi \,, \tag{4.13}$$

where again $\varphi = \theta_2 - \theta_1$.

Thus we have arrived at the same simple relation between the supercurrent and the phase difference of the superconducting wavefunctions in the junction electrodes as before. In the following we shall always assume that this relation holds.

4.3 The AC Josephson Effect

So far we have considered the case of a small current $I < I_c$ through a weak link. If the current supplied by an external source exceeds the critical value I_c, it causes a voltage V to appear across the junction. What consequences should then be expected?

The behavior of a quantum-mechanical system is described by the Schrödinger equation

$$\mathrm{i}\hbar \frac{\partial \psi}{\partial t} = \hat{H}\psi \,, \tag{4.14}$$

where $\hat{H}$ is the system's Hamiltonian.

The wavefunction of a stationary state ψ_1 satisfies the equation

$$\hat{H}\psi_1 = E\psi_1 \,,$$

where E is the energy of the state and $\psi_1 = |\psi_1|\,\mathrm{e}^{\mathrm{i}\theta(t)}$, with $|\psi_1|$ being independent of time. On substitution of this wavefunction into (4.14) we get

$$-\hbar \frac{\partial \theta}{\partial t} = E . \tag{4.15}$$

The presence of the voltage V across the weak link suggests that the Cooper pair energies in superconductors on either side of the junction, E_1 and E_2, are related to each other by

$$E_1 - E_2 = 2eV , \tag{4.16}$$

because the charge of a pair is $2e$.

Substituting (4.16) into (4.15), we obtain the second fundamental Josephson relation:

$$2eV = \hbar \frac{\partial \varphi}{\partial t} . \tag{4.17}$$

The same relation can be obtained by subtracting (4.7) from (4.6).

Now, what happens to a Josephson junction if we apply a constant current $I > I_c$ to it? Since the supercurrent cannot exceed I_c, it is obvious that a current of normal electrons I_n must start flowing through the junction, in addition to the supercurrent. This conclusion leads us directly to the so-called resistively shunted model of the Josephson junction (RSJ) in which the latter is considered as a circuit made up of the Josephson junction itself and a normal resistance connected in parallel (Fig. 4.5).

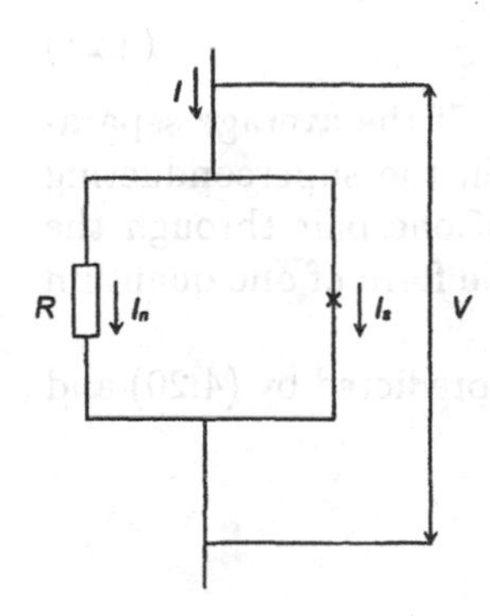

Fig. 4.5. Resistively shunted model of a Josephson junction. The supercurrent through the junction is $I_s = I_c \sin \varphi$

The total current I is then a sum of the normal current V/R and the supercurrent $I_s = I_c \sin \varphi$:

$$I = I_c \sin \varphi + \frac{\hbar}{2eR} \frac{\partial \varphi}{\partial t} , \tag{4.18}$$

where R is the normal-state resistance of the junction. This differential equation can be easily integrated. Substituting the solution into (4.17), we obtain the voltage across the junction as

$$V(t) = R \frac{I^2 - {I_c}^2}{I + I_c \cos \omega t} , \tag{4.19}$$

$$\omega = \frac{2e}{\hbar} R \sqrt{I^2 - I_c{}^2} . \tag{4.20}$$

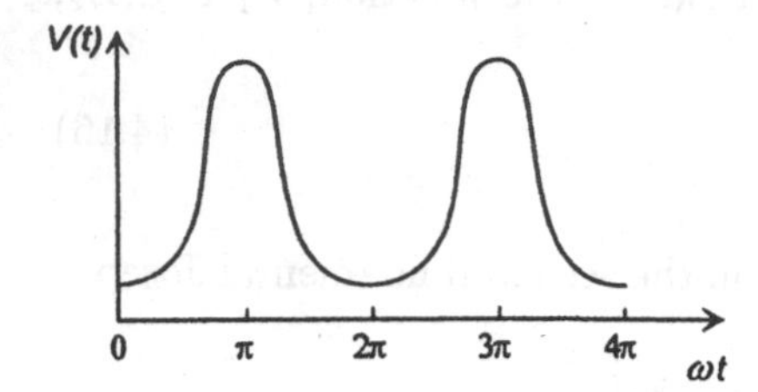

Fig. 4.6. Voltage across the Josephson junction at $I \geq I_c$

Thus we have found a fascinating property of the Josephson junction. If an external dc current I through the junction exceeds the critical current I_c, it causes a voltage V to appear across the junction, which oscillates periodically with time. This phenomenon is often referred to as Josephson radiation. The time dependence of V is sketched in Fig. 4.6. The frequency of the ac voltage depends on the amount by which the current through the junction, I, exceeds the critical value I_c (see (4.20)).

If we now connect the junction to a dc voltmeter, the latter will register the average value over one period, $\overline{V}$. Averaging of (4.19) over time gives

$$2e\overline{V} = \hbar\omega . \tag{4.21}$$

This formula has a very clear physical interpretation. If the average separation between the energy levels of the Cooper pairs in the superconducting electrodes is $2e\overline{V}$, then, as a result of the transfer of one pair through the weak link, this energy is released by the junction in the form of one quantum of electromagnetic radiation.

The current–voltage characteristic of the junction predicted by (4.20) and (4.21) is shown in Fig. 4.7.

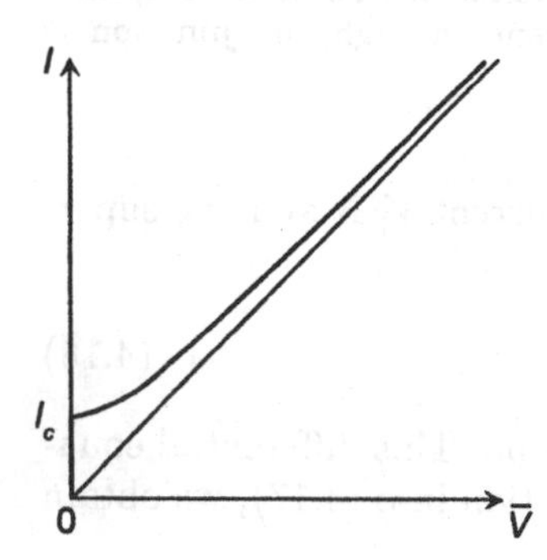

Fig. 4.7. Current–voltage characteristic of a Josephson junction

Note that the relations (4.17) and (4.21) are absolutely precise and fundamental. Their precision has been verified with remarkable accuracy by numerous experiments. The first experimental observation of the Josephson radiation was reported in 1964 by Yanson, Svistunov and Dmitrenko [11].

4.4 Response of a Josephson Junction to an External Magnetic Field

In this section we shall restrict our attention to Josephson junctions without concentration of current, that is, to sandwiches and tunnel junctions. Suppose, for example, that such a junction consists of two bulk superconducting slabs separated by a thin insulating layer, as in Fig. 4.8. If this system is placed in an external magnetic field parallel to the plane of the junction, a screening supercurrent will be generated at the outer surfaces of the slabs. This current circulates within a surface layer of thickness λ and, in doing so, has to cross the weak link, where the critical current density is very small. Therefore, in order to maintain its dissipation-free flow, the current must spread over a rather wide area protruding into the junction. This situation is illustrated in Fig. 4.8. Let us try to describe it in mathematical terms.

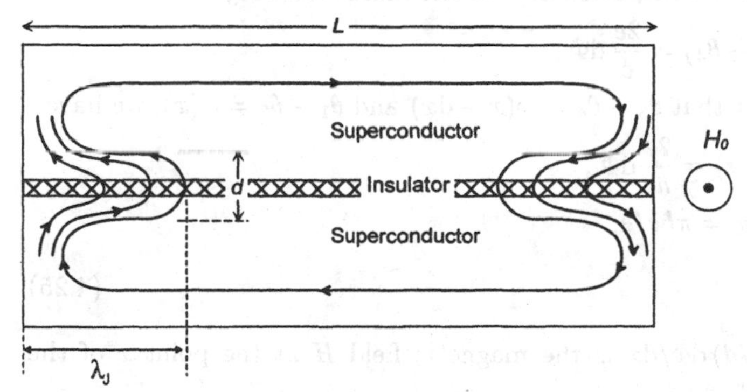

Fig. 4.8. Tunnel Josephson junction in a magnetic field H_0. The distribution of the screening (Meissner) supercurrent is shown by lines with arrows

4.4.1 The Ferrell–Prange Equation

Consider a Josephson junction in an external magnetic field. The plane of the junction coincides with the x axis and the magnetic field is directed along the z axis. The length along the y axis of the region penetrated by the current and the magnetic field is $d = 2\lambda + t$ (see Fig. 4.8). Here t is the thickness of the insulating layer. Consider two closely spaced pairs of points in the vicinity of the junction: *1, 2* and *3, 4*, as in Fig. 4.9. All the points are chosen outside the area penetrated by the magnetic field and the distance between them is taken to be $\mathrm{d}x$.

The generalized momentum of a Cooper pair has the form

$$\hbar\nabla\theta = 2m\boldsymbol{v}_s + \frac{2e}{c}\boldsymbol{A}\,, \tag{4.22}$$

where θ is the phase of the wavefunction, $\boldsymbol{v}_s$ the pair velocity, m the electron mass, e the electron charge, and $\boldsymbol{A}$ the vector potential. Let us integrate this equation along the dashed curves shown in Fig. 4.9. Since both curves are outside the area penetrated by the supercurrent, we have $v_s = 0$:

$$\hbar\left[\int_1^3 \nabla\theta\cdot d\boldsymbol{l} + \int_4^2 \nabla\theta\cdot d\boldsymbol{l}\right] = \frac{2e}{c}\left[\int_1^3 \boldsymbol{A}\cdot d\boldsymbol{l} + \int_4^2 \boldsymbol{A}\cdot d\boldsymbol{l}\right]\,. \tag{4.23}$$

The distance d is assumed to be negligibly small. Then the right-hand side of (4.23) can be approximated by

$$\frac{2e}{c}\oint \boldsymbol{A}\cdot d\boldsymbol{l} = \frac{2e}{c}\,d\Phi\,, \tag{4.24}$$

where $d\Phi$ is the magnetic flux enclosed by the contour comprising the dashed curves *1–3* and *4–2* which are supplemented with missing sections *3–4* and *2–1*. Carrying out the integration on the left-hand side we get

$$\hbar(\theta_3 - \theta_1 + \theta_2 - \theta_4) = \frac{2e}{c}\,d\Phi\,.$$

Taking into account that $\theta_3 - \theta_4 = \varphi(x + dx)$ and $\theta_1 - \theta_2 = \varphi(x)$, we have

$$\varphi(x + dx) - \varphi(x) = \frac{2e}{\hbar c}\,d\Phi\,,$$

or, recalling that $\Phi_0 = \pi\hbar c/e$,

$$\frac{d\varphi}{dx} = \frac{2\pi}{\Phi_0}\frac{d\Phi}{dx}\,. \tag{4.25}$$

Noting that $(1/d)\,d\Phi/dx$ is the magnetic field H at the point x of the junction, we can write

$$H = \frac{\Phi_0}{2\pi d}\frac{d\varphi}{dx}\,. \tag{4.26}$$

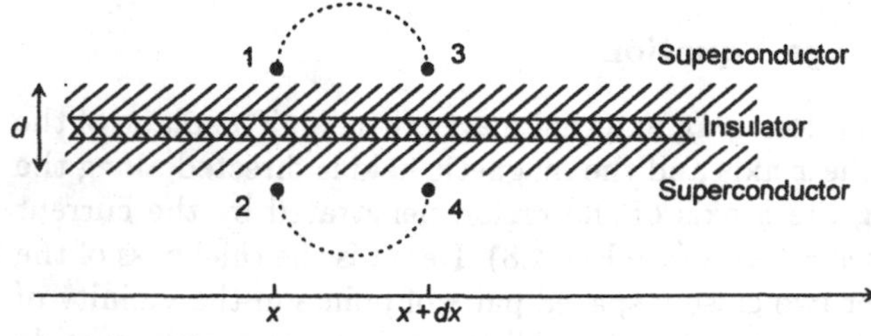

Fig. 4.9. Section near the edge of the Josephson junction. The *dashed area* is the region penetrated by the magnetic field

This magnetic field H is directed along the z axis. By Maxwell's equations, the supercurrent through the junction in the direction of the y axis is

$$j_s = \frac{c}{4\pi}\frac{dH}{dx}.$$

Inserting (4.26) and recalling the fundamental Josephson relation between the current and the phase difference, $j_s = j_c \sin\varphi$, we arrive at the so-called Ferrell–Prange equation [36]:

$$\frac{d^2\varphi}{dx^2} = \frac{1}{\lambda_J^2}\sin\varphi\,, \tag{4.27}$$

where λ_J is

$$\lambda_J = \left(\frac{c\Phi_0}{8\pi^2 j_c d}\right)^{1/2}. \tag{4.28}$$

The solution of (4.27), $\varphi(x)$, gives the phase difference distribution over the junction. Consider the case of a very weak external field $H_0 \ll \Phi_0/(2\pi\lambda_J d)$. Both the currents through the junction and the phase difference φ are then small. Hence we can rewrite (4.27) as

$$\frac{d^2\varphi}{dx^2} = \frac{1}{\lambda_J^2}\varphi\,.$$

This equation can be easily solved to give

$$\varphi(x) = \varphi(0)\exp(-x/\lambda_J)\,. \tag{4.29}$$

Substituting this solution into (4.26), we find the magnetic field in the junction:

$$H(x) = H_0 \exp(-x/\lambda_J)\,.$$

One can see that the quantity λ_J, which has the dimensions of length, represents the depth of magnetic field penetration into the Josephson junction. λ_J is usually referred to as the Josephson penetration depth.

In the International System of Units (SI), equation (4.28) takes the form:

$$\lambda_J = \left(\frac{\Phi_0}{2\pi\mu_0 j_c d}\right)^{1/2},$$

where $\mu_0 = 4\pi \times 10^{-7}\ \mathrm{H\,m^{-1}}$, $\Phi_0 = 2.07 \times 10^{-15}$ Wb, j_c is the density of the critical current through the junction in $\mathrm{A\,m^{-2}}$, and d is taken in m. Taking parameters typical of a tunnel Josephson junction: $d \sim 10^{-5}$ cm, $j_c \sim 10^2\ \mathrm{A\,cm^{-2}}$, we obtain $\lambda_J \sim 0.1$ mm, i.e., it is a macroscopic length.

4.4.2 Magnetic Field Penetration into a Josephson Junction. Josephson Vortices

In the previous section we examined the behavior of a Josephson junction in a weak external field which penetrates the junction to the depth λ_J (the Josephson penetration depth). Now, what happens if the external field increases? It turned out that the behavior of a Josephson junction in an external magnetic field is reminiscent of the behavior of a type-II superconductor. Just as with type-II superconductors, when the external field exceeds a certain critical value H_{c1}, which is a characteristic of the junction, the field starts to penetrate into the junction in the form of superconducting vortices, each carrying one magnetic flux quantum Φ_0. In the case of a Josephson junction, they are called Josephson vortices.

Indeed, one of the solutions of the Ferrell–Prange equation (4.27) has the form

$$\varphi_0(x) = 4 \arctan \left[\exp(x/\lambda_J)\right] . \tag{4.30}$$

One can easily verify that this solution satisfies (4.27). The functions $\varphi_0(x)$, $d\varphi_0/dx \sim H$, and $d^2\varphi_0/dx^2 \sim j_s$ are shown in Fig. 4.10. A Josephson vortex represents a so-called soliton, that is, it is a localized solitary excitation of an extended Josephson junction.[4] The size of a static vortex along the x axis of the junction is $\sim 2\lambda_J$ and along the y axis $\sim d \ll 2\lambda_J$.

Thus, beginning at the field H_{c1}, penetration of Josephson vortices into the junction becomes energetically favorable. Having penetrated into the junction, the vortices form a linear chain and the junction goes into the mixed state. So far we have a complete analogy with a type-II superconductor. But is the analogy really complete? The answer is 'no'.

The thing is that, unlike an Abrikosov vortex (see Chap. 5), a Josephson vortex does not have a normal core. The existence of the normal core in type-II superconductors relates to the second critical field H_{c2}: Superconductivity disappears when an external magnetic field presses the vortices so close together that their normal cores come into contact. The absence of a normal core for Josephson vortices implies that there is no upper critical field for a Josephson junction. Nevertheless, as we shall soon find out, the dependence of the maximum current through the junction on magnetic field can be rather curious.

Let us find the lower critical field of the junction, H_{c1}. In order to do that, we shall analyze the free energy of the junction.

As the current through the junction increases from zero to a certain value j_s over a time t, a certain amount of energy is being stored by the junction. Per unit area of the junction, this energy is

[4] *Editors' note.* Here the author discusses only static properties of Josephson vortices, also known as Josephson fluxons. When a dc bias current is applied to a Josephson junction, the vortices can move along the junction and their motion leads to the appearance of a dc voltage accompanied by Josephson radiation, according to (4.21). This dynamic state is described by the so-called perturbed sine-Gordon equation [37], which is a more general case of (4.27) for the space- and time-dependent φ. Moving Josephson fluxons exhibit relativistic properties of solitons (nonlinear solitary electromagnetic waves) and have been extensively studied both theoretically and experimentally. For reviews on solitons in Josephson junctions, we refer the reader to [38, 39].

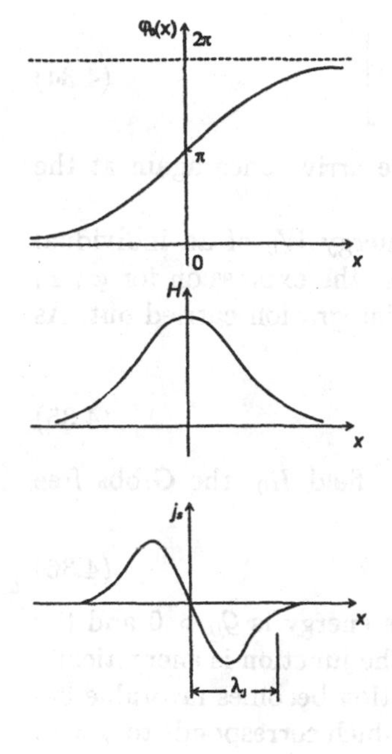

Fig. 4.10. Phase difference $\varphi_0(x)$, magnetic field $H(x)$, and supercurrent $j_s(x)$ for a Josephson vortex

$$w_J = \int_0^t j_s V \, dt \, . \tag{4.31}$$

Here V is the voltage across the junction, which appears during the process of increasing the current and obeys the general relation (4.17). Substituting here $j_s = j_c \sin\varphi$ and (4.17), we obtain, after carrying out elementary integration,

$$w_J = \frac{\hbar}{2e} j_c (1 - \cos\varphi) \, . \tag{4.32}$$

One notices that the Josephson junction plays the role of a nonlinear inductance which stores the energy. In addition to this energy, there is also a magnetic energy, which, per unit area of the junction, is given by

$$w_H = \frac{H^2}{8\pi} d \, . \tag{4.33}$$

Taking the sum of (4.32) and (4.33) and carrying out integration over the length of the junction L (see Fig. 4.8), we obtain the total free energy of the junction, per unit length along the direction of the magnetic field (i.e., the z direction):

$$W = \int_0^L dx \left[\frac{H^2}{8\pi} d + \frac{\hbar}{2e} j_c (1 - \cos\varphi) \right] \, .$$

Substituting (4.26) into this equation, we can modify the expression for W to

$$W = \int_0^L \mathrm{d}x \left[\frac{\Phi_0{}^2}{32\pi^3 d} \left(\frac{\mathrm{d}\varphi}{\mathrm{d}x} \right)^2 + \frac{\hbar}{2e} j_c (1 - \cos\varphi) \right] . \tag{4.34}$$

Using the variational method to minimize W, we arrive once again at the Ferrell–Prange equation (4.27).

From (4.34), one can easily derive the free energy W_0 of an individual Josephson vortex in an infinite junction. To do so, the expression for $\varphi_0(x)$ (4.30) should be substituted into (4.34) and the integration carried out. As a result we get

$$W_0 = \frac{4\Phi_0 j_c}{\pi c} \lambda_J . \tag{4.35}$$

If the junction is placed in an external magnetic field H_0, the Gibbs free energy of a Josephson vortex becomes, by (1.22),

$$\mathcal{G}_0 = W_0 - \Phi_0 H_0 / 4\pi . \tag{4.36}$$

Thus, if the external field is sufficiently weak, the energy is $\mathcal{G}_0 > 0$ and the existence of a Josephson vortex in the interior of the junction is energetically unfavorable. Penetration of a vortex into the junction becomes favorable beginning at a certain value of the external field H_0 which corresponds to $\mathcal{G} = 0$. This field is called the lower critical field H_{c1}:

$$H_{c1} = \frac{2}{\pi^2} \frac{\Phi_0}{\lambda_J d} . \tag{4.37}$$

The field H_{c1} is obviously less than the field at the center of the vortex, $H(0)$. The latter, from (4.30) and (4.26), is

$$H(0) = \frac{\Phi_0}{\pi \lambda_J d} . \tag{4.38}$$

The field $H(0)$ is effectively the superheating field for the Meissner state of the junction.

4.4.3 Maximum Zero-Voltage Current of a Josephson Junction. High Magnetic Fields

Let us turn to high magnetic fields:

$$H_0 \gg \frac{\Phi_0}{2\pi \lambda_J d} . \tag{4.39}$$

We assume that the size of the junction is $L \ll 2\lambda_J$. Thus we can ignore the self-induced field due to the current passing through the junction, because it is much less than H_0. The inequality (4.39) can be interpreted as a condition on the distance between the neighboring Josephson vortices in the chain: their distance must be much less than λ_J. Then it follows from (4.26) that

$$\frac{\mathrm{d}\varphi}{\mathrm{d}x} = \frac{2\pi d}{\Phi_0} H_0$$

and, carrying out integration, we get

$$\varphi(x) = \frac{2\pi d}{\Phi_0} H_0 x + C\,, \tag{4.40}$$

where C is an arbitrary integration constant. Substituting (4.40) into the first Josephson relation (4.13) we obtain

$$j_s = j_c \sin\left(\frac{2\pi x}{a} + C\right)\,, \tag{4.41}$$

where we use the notation

$$a = \Phi_0 / H_0 d\,. \tag{4.42}$$

It follows from (4.41) that the chain of Josephson vortices in this limit is indeed very dense. It is sketched in Fig. 4.11, together with the corresponding distribution of the tunneling supercurrent.

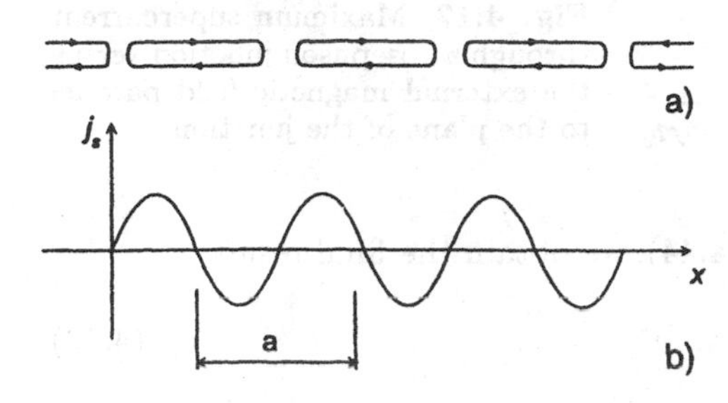

Fig. 4.11. (a) Densely packed chain of vortices representing the mixed state in a tunnel Josephson junction; **(b)** corresponding distribution of the tunneling supercurrent through the junction

The total current through the junction can be found by integration of j_s in (4.41) over the entire length of the junction:

$$I_s = j_c \int_{-L/2}^{L/2} \sin(2\pi x/a + C)\,dx\,.$$

As a result we have:

$$I_s = j_c L \frac{\sin(\pi L/a)}{\pi L/a} \sin C\,. \tag{4.43}$$

It follows from (4.43) that, in a fixed magnetic field, a change of the total current through the junction, I_s (which is supplied by an external source) causes a subsequent change of the constant C. The latter adjusts itself to each value of I_s. Then the maximum total current, I_{max}, which can pass through the junction without dissipation is simply the amplitude of the quantity written in (4.43) before $\sin C$:

$$I_{max} = I_c \left|\frac{\sin(\pi L/a)}{\pi L/a}\right|\,, \tag{4.44}$$

with the notation

$$I_c = j_c L\ .$$

The expression for I_{max}, equation (4.44), can be written in a more convenient form if we note that, according to (4.42),

$$\pi L/a = \pi\Phi/\Phi_0\ , \tag{4.45}$$

where Φ is the total magnetic flux through the Josephson junction:

$$\Phi = H_0 L d\ .$$

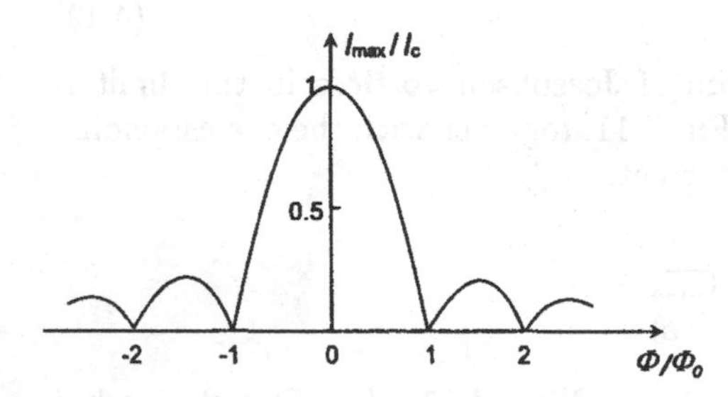

Fig. 4.12. Maximum supercurrent through a Josephson junction versus the external magnetic field parallel to the plane of the junction

On substitution of (4.45) into (4.44), we obtain the final result:

$$I_{max} = I_c \left| \frac{\sin(\pi\Phi/\Phi_0)}{\pi\Phi/\Phi_0} \right| . \tag{4.46}$$

The dependence of I_{max} on the external magnetic field $H_0 = \Phi/(Ld)$ is shown in Fig. 4.12. As one can conclude from the plot and from (4.46), the mixed state in the junction is absolutely unstable (that is, a negligibly small current suffices to destroy it) when the total magnetic flux Φ is equal to an integral number of Φ_0. In contrast, it is most stable whenever there are a

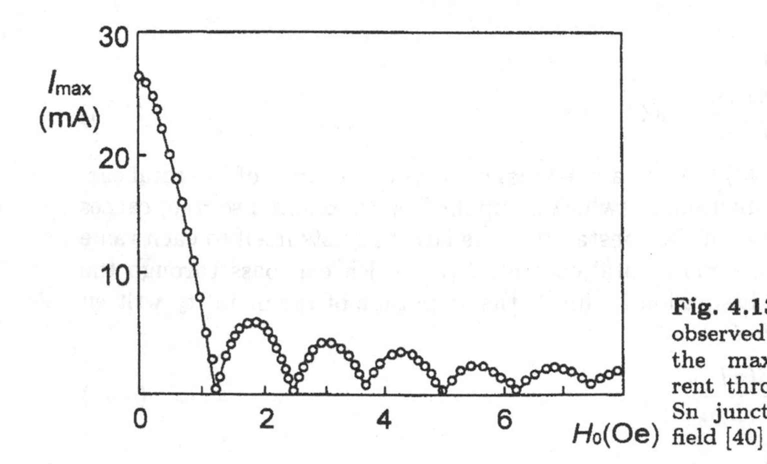

Fig. 4.13. Experimentally observed dependence of the maximum supercurrent through a Sn–SnO_x–Sn junction on magnetic field [40]

half-integral number of flux quanta enclosed in the junction. Experiment [40] has given a spectacular verification of (4.46), as one can see from Fig. 4.13.

4.5 Superconducting Quantum Interferometers

This section is concerned with the basic principles underlying the devices known as superconducting quantum interference devices (SQUIDs). These devices, though very simple in design, have opened new horizons in low-temperature measurement techniques. Many SQUID-based instruments are unique in their sensitivity. The most celebrated examples are SQUID magnetometers, which are able to resolve flux increments of $\sim 10^{-10}$ G, and precision voltmeters with the sensitivity of $\sim 10^{-15}$ V.

So, what is SQUID?

There are two basic types of SQUIDs to distinguish: a two-junction dc SQUID and a single-junction rf SQUID.

4.5.1 The Two-Junction SQUID (DC SQUID)

This device consists of two Josephson junctions connected in parallel. In practice, the circuit consists of two bulk superconductors which, together with the Josephson junctions a and b, form a ring as in Fig. 4.14. The flux through the loop of the SQUID is generated by a magnetic coil placed in the interior of the ring. To understand how this type of SQUID operates, we need to know how the maximum zero-voltage current I_{max} through the device depends on the total magnetic flux Φ enclosed in the SQUID loop.

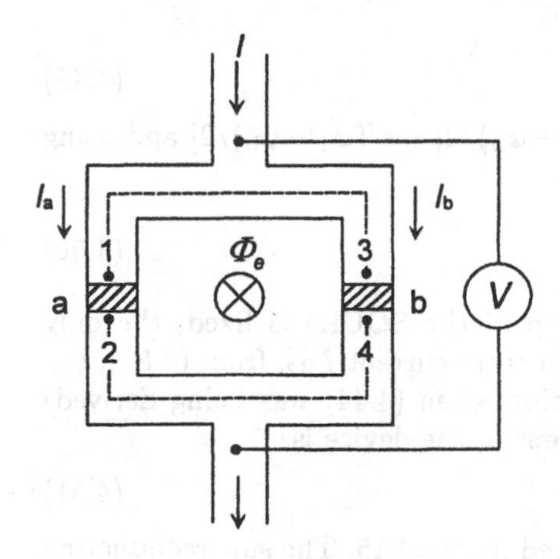

Fig. 4.14. Superconducting interferometer: two Josephson junctions, a and b, are connected in parallel. The interior of the SQUID loop is threaded by a magnetic flux Φ

Consider two pairs of points in the interiors of the superconductors: (*1, 2*) and (*3, 4*), all of which are close to the junctions a and b, as illustrated in

Fig. 4.14. Carrying out integration of (4.22) along the dashed contour from point *1* to point *3* and from point *4* to point *2*, we obtain[5]

$$\theta_3 - \theta_1 + \theta_2 - \theta_4 = \frac{2e}{\hbar}\left(\int_1^3 \boldsymbol{A}\cdot \mathrm{d}\boldsymbol{l} + \int_4^2 \boldsymbol{A}\cdot \mathrm{d}\boldsymbol{l}\right) . \tag{4.47}$$

The term $2m\boldsymbol{v}_s$ has been omitted because the contour passes everywhere through the interior of the superconductor, well away from the edges. There is no supercurrent there and $\boldsymbol{v}_s = 0$. The distance between points *1* and *2*, as well as between *3* and *4*, is small compared to the length of the dashed contour. In addition, the vector potential $\boldsymbol{A}$ does not have any special features near the junctions. Therefore, the right-hand side of (4.47) can be supplemented by an integral along the sections *3–4* and *1–2*. As a result we get

$$\varphi_a - \varphi_b = \frac{2e}{\hbar}\oint \boldsymbol{A}\cdot \mathrm{d}\boldsymbol{l} ,$$

or

$$\varphi_a - \varphi_b = 2\pi\Phi/\Phi_0 , \tag{4.48}$$

where Φ is the total magnetic flux enclosed in the loop of the interferometer, $\varphi_a = \theta_2 - \theta_1$, $\varphi_b = \theta_4 - \theta_3$, and $\Phi_0 = \pi\hbar c/e$ is the magnetic flux quantum.

The current through the junction a is, from (4.1),

$$I_a = I_c \sin\varphi_a ,$$

and through the junction b,

$$I_b = I_c \sin\varphi_b$$

(the junctions are assumed to be identical and characterized by the same value of the critical current, I_c). The total current through the interferometer is then the sum of I_a and I_b:

$$I = I_c\,(\sin\varphi_a + \sin\varphi_b) . \tag{4.49}$$

Noting that $\sin\varphi_a + \sin\varphi_b = 2\sin[(\varphi_a + \varphi_b)/2]\cos[(\varphi_a - \varphi_b)/2]$ and using (4.48), we can modify (4.49) to the form

$$I = 2I_c \cos\frac{\pi\Phi}{\Phi_0}\sin\left(\varphi_b + \frac{\pi\Phi}{\Phi_0}\right) . \tag{4.50}$$

If the total flux enclosed in the loop of the SQUID is fixed, the only parameter which adjusts itself to a given total current I is, from (4.50), φ_b. It follows (just as in the preceding section when (4.44) was being derived) that the maximum dissipation-free current of the device is

$$I_{max} = 2I_c\,|\cos(\pi\Phi/\Phi_0)| . \tag{4.51}$$

The dependence of I_{max} on Φ is illustrated in Fig. 4.15. The superconducting state of the ring is most stable with respect to the external current I when an

[5] Throughout this section, the International System of Units (SI) is employed.

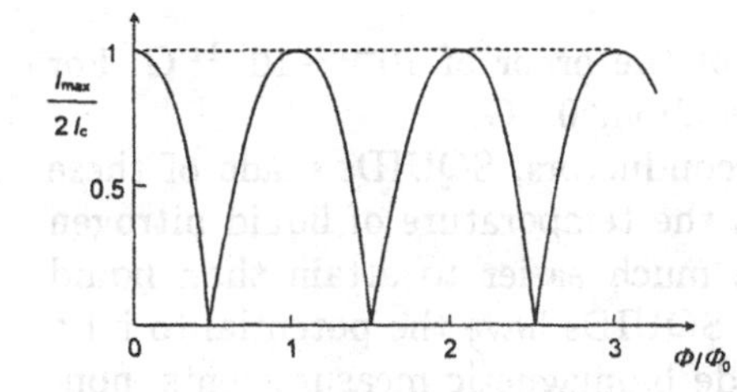

Fig. 4.15. Dependence of the maximum supercurrent through the two-junction interferometer on the total magnetic flux through its interior

integral number of flux quanta are enclosed in the interferometer. In contrast, a half-integral number of flux quanta in the loop corresponds to an unstable superconducting state. That is to say, a negligibly small current I suffices in the latter case to drive the device to the resistive state, with a finite voltage across the junction (see Fig. 4.14).

We would like to emphasize that Φ is the total flux through the loop of the interferometer. The flux supplied externally by the magnetic coil, Φ_e, is related to Φ by

$$\Phi = \Phi_e - LI_{sc} \,,$$

where L is the inductance of the interferometer and I_{sc} is the screening current circulating in it. The critical current of the SQUID is periodic in Φ_e, too, with period Φ_0. This dependence is illustrated in Fig. 4.16.

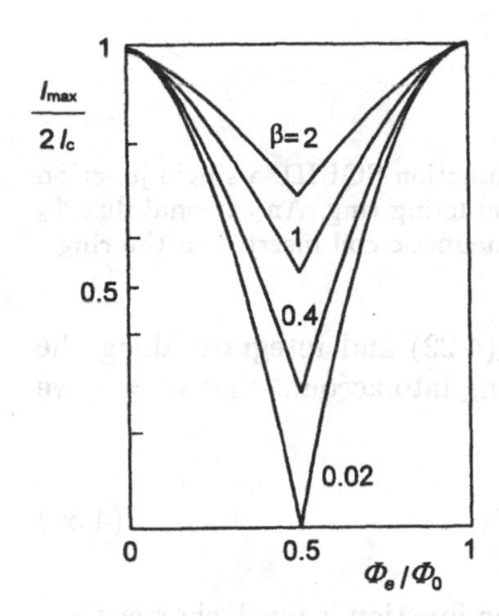

Fig. 4.16. Critical current of the two-junction SQUID versus external magnetic flux, for several values of the parameter $\beta = 2LI_c/\Phi_0$ [41]

A detailed analysis of the two-junction SQUID operation is given by Clarke in [41].

The unique sensitivity of dc SQUIDs can be employed in many situations where a variation of the quantity of interest can be converted into a variation of magnetic flux. The only limit on it is set by thermal noise in Josephson junctions which is estimated as several units per $10^{-5}\,\Phi_0$, provided the time of the measurements is longer than 1 s. Therefore, in principle, it is possible to resolve increments of magnetic field of the order of 10^{-10}–10^{-11} G. For comparison, the Earth's magnetic field is about 0.5 G.

After the discovery of high-T_c superconductors, SQUIDs made of these materials were shown to operate well at the temperature of liquid nitrogen (77 K). Since this temperature range is much easier to attain than liquid helium temperature (4.2 K), the high-T_c SQUIDs have the potential to fill a broad range of applications. These include biomagnetic measurements, non-destructive evaluation, and geomagnetic probing. The present status of high-T_c SQUIDs is described in a recent review [42].

4.5.2 The Single-Junction SQUID (RF SQUID)

The basic element of a single-junction SQUID is a superconducting ring containing a Josephson junction. Consider two points, *1* and *2*, in the vicinity of the junction, as shown in Fig. 4.17. The dashed contour *1–2* passes through the interior of the superconductor in such a way that its distance from the edges is everywhere larger than λ. Therefore, there is no supercurrent at any point of the contour and $\boldsymbol{v}_s = 0$.

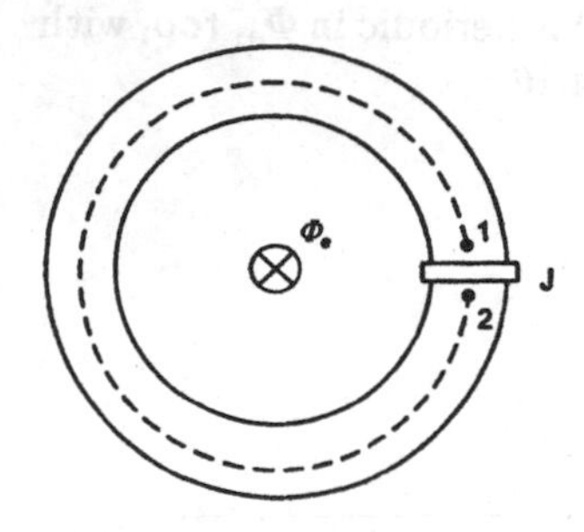

Fig. 4.17. Single-junction SQUID: a single junction J closes a superconducting ring. An external flux Φ_e is generated by a magnetic coil inserted in the ring

Let us go back to the general equation (4.22) and integrate along the dashed contour from point *1* to point *2*. Taking into account that $v_s = 0$, we have

$$\hbar(\theta_2 - \theta_1) = 2e \int_1^2 \boldsymbol{A} \cdot \mathrm{d}\boldsymbol{l} \,. \tag{4.52}$$

The distance between points *1* and *2* across the junction is much shorter than their distance along the dashed contour, and the vector potential does not have peculiarities in the vicinity of the junction. Therefore, the variation of the right-hand side of (4.52) will be negligible if we supplement it with an integral $\int_2^1 \boldsymbol{A} \cdot \mathrm{d}\boldsymbol{l}$ across the junction. Then (4.52) can be rewritten as

$$\hbar\varphi = 2e \oint \boldsymbol{A} \cdot \mathrm{d}\boldsymbol{l} \,, \tag{4.53}$$

where φ is the phase difference across the junction J. Equation (4.53) can be further reduced to

$$\varphi = 2\pi\Phi/\Phi_0 \,, \tag{4.54}$$

where Φ is the total magnetic flux enclosed in the SQUID loop. In general, this flux Φ is not equal to the externally supplied flux Φ_e. Their difference is due to the screening current circulating in the superconducting ring:

$$\Phi = \Phi_e - LI_{sc} \,, \tag{4.55}$$

where L is the inductance of the ring. Since the current I_{sc} passes through both the ring and the junction, its relation to the phase difference of the superconducting electron wavefunction is given by the well-known expression (4.1). Together with (4.54) and (4.55), this yields

$$\Phi_e = \Phi + LI_c \sin(2\pi\Phi/\Phi_0) \,. \tag{4.56}$$

This formula can be considered as an implicit relation between Φ and Φ_e. It is illustrated graphically in Fig. 4.18.

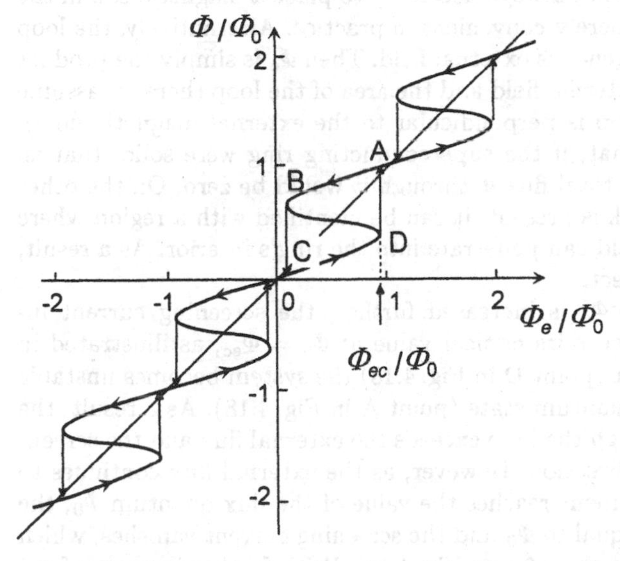

Fig. 4.18. Total magnetic flux Φ enclosed in the loop of a single-junction interferometer versus the external flux Φ_e through the loop

Let us now analyze the physics of SQUID operation. First we increase the external flux Φ_e (that is, we increase the current through the magnetic coil inserted in the SQUID loop). As an immediate response, a screening current I_{sc} starts to flow in the ring and generates its own magnetic flux which partially cancels the external flux Φ_e. Therefore, the total flux is less than Φ_e. Why is the cancellation only partial? If the ring were solid, without the weak link, the cancellation would be complete and the total flux would be $\Phi = 0$.

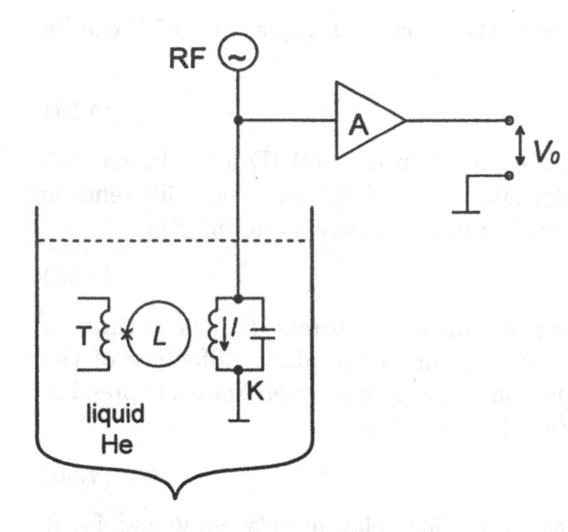

Fig. 4.19. Block-diagram of an rf SQUID. RF is a high-frequency current supply (of large internal resistance), A a high-frequency amplifier, L a single-junction interferometer, T a secondary coil of a flux transformer, and K a high-frequency resonant circuit

This situation can also be described in a different way. In order to supply the external flux Φ_e, it is not always necessary to place a magnetic coil in the loop of the device; it is merely convenient in practice. Alternatively, the loop can be placed in a homogeneous external field. Then Φ_e is simply the product of the induction of the external field and the area of the loop (here we assume that the plane of the loop is perpendicular to the external magnetic field). Then it is easy to see that, if the superconducting ring were solid, that is, without a weak link, the total flux Φ through it would be zero. On the other hand, when the weak link is present, it can be identified with a region where the external magnetic field can penetrate into the ring's interior. As a result, the screening is not perfect.

As the external flux Φ_e is increased further, the screening current increases, too, until it reaches its critical value at $\Phi_e = \Phi_{ec}$, as illustrated in Fig. 4.18. At this moment (point D in Fig. 4.18) the system becomes unstable and jumps to the next quantum state (point A in Fig. 4.18). As a result, the total magnetic flux through the loop exceeds the external flux and the screening current changes its direction. However, as the external flux continues to increase and at some moment reaches the value of the flux quantum Φ_0, the total flux also becomes equal to Φ_0 and the screening current vanishes, which is effectively equivalent to $\Phi_e = 0$ (see Fig. 4.18). With further increase of Φ_e, this switching process repeats itself with period Φ_0. If Φ_e is decreased, the jumps will start from points analogous to point B. Thus, a cyclic variation of the external flux Φ_e is accompanied by a hysteresis loop CDABC. The area of the loop is proportional to the energy dissipated in the junction.

Now our knowledge of the properties of a superconducting ring with a single Josephson junction is sufficient to understand how the rf SQUID operates. Its block-diagram is shown in Fig. 4.19. A current of frequency ν is supplied by a high-impedance ac current source to the resonant circuit K.

(Frequencies $\nu \approx$ 10–20 MHz are used most often. However, there are also SQUIDs operating at higher frequencies such as several GHz). The coil of the resonant circuit is coupled to the loop of the interferometer I. The rf voltage across the circuit K is amplified by the amplifier A. This voltage, V_0, is the output of the device. The device input is the coil T inductively coupled to the SQUID loop L.

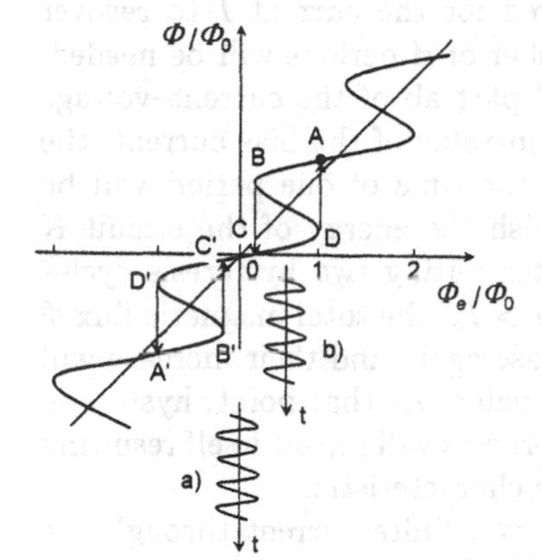

Fig. 4.20. Variation of the external flux with time: (a) when zero flux is generated by the input coil T ($\Phi_T = 0$); (b) when $\Phi_T = \Phi_0/2$

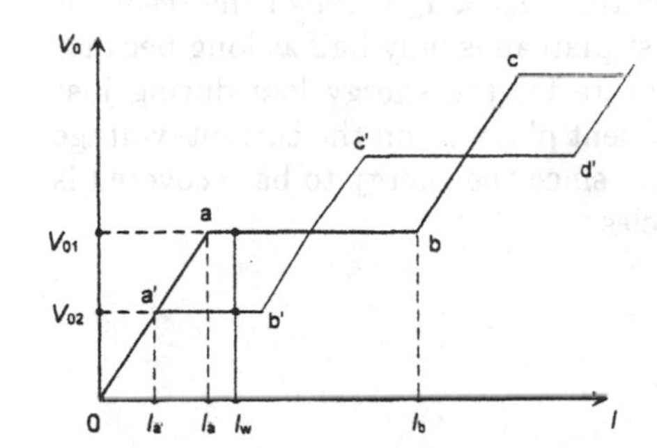

Fig. 4.21. RF current–voltage characteristic of the SQUID; bold line (0abc...) corresponds to the input flux $\Phi_T = 0$, thin line (0a′b′c′d′...) – to $\Phi_T = \Phi_0/2$

Now let us work out how the output V_0 relates to the value of the current I through the resonant circuit K. Assume that the current through the input coil T is zero. Then the external flux Φ_e is generated entirely by the circuit K and has the frequency ν, as in Fig. 4.20 (a). In turn, the sinusoidal variation of Φ_e causes a sinusoidal variation of the total flux Φ (see Fig. 4.20) resulting in an rf voltage V_0 at the output. An increase of the bias current I causes a proportional increase of the amplitude of Φ_e and, as a consequence, of Φ and V_0. Thus the dependence of V_0 on I is linear (at the initial part 0a of the rf current–voltage characteristic in Fig. 4.21) until the current reaches a value such that $\Phi_e = \Phi_{ec}$ (see Fig. 4.18). At this moment, a discontinuous jump of the total flux occurs, corresponding to closing a hysteresis loop on the $\Phi(\Phi_e)$ curve, and an amount of heat proportional to the area of the loop is dissipated. The necessary amount of energy, which is quite large, is borrowed from the resonant circuit K. As a result, the amplitude of the current I sharply decreases and it takes a long time (a large number of rf oscillation periods) for the current to recover its initial value. A further increase of the amplitude of current supplied by the generator will not result in an increase of the output voltage V_0 because the above process will repeat itself again and again. The only difference will be that the time required for the current I to recover will steadily decrease, that is, a smaller number of rf periods will be needed. This means that we arrive at the horizontal part ab of the current–voltage characteristic in Fig. 4.21. Finally, at a certain value of the bias current, the current I will reach a value (I_b) such that the time of one period will be sufficient for the rf current supply to replenish the energy of the circuit K that has been absorbed by the interferometer during two hysteresis cycles DABC and D'A'B'C' shown in Fig. 4.20. At $I > I_b$, the total magnetic flux Φ and the output voltage V_0 will start to increase again and their increase will continue until Φ_e reaches the second critical value. At that point, hysteretic energy losses will occur again and the whole process will repeat itself resulting in the second plateau on the current–voltage characteristic.

Let us now turn to the case when there is a finite current through the input coil T generating a dc flux $\Phi_T = \Phi_0/2$. The time dependence of Φ_e in this case is illustrated in Fig. 4.20 (b). Clearly, the critical value Φ_{ec} is now reached at a significantly smaller current $I_{a'} < I_a$ through the resonant circuit. In addition, the length of the first plateau is only half as long because the current supply now has to compensate for the energy loss during just one hysteresis cycle. However, all subsequent plateaus on the current–voltage characteristic are of normal length again, since the energy to be recovered is dissipated during pairs of hysteresis cycles.

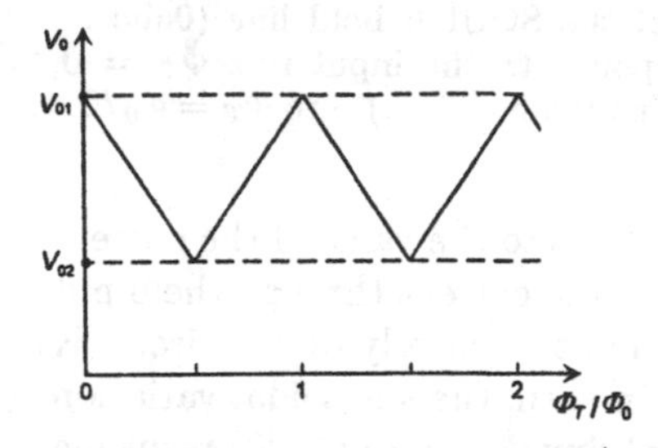

Fig. 4.22. Dependence of the output rf voltage V_0 on the external flux Φ_T generated by the input coil T

As the final stage of our analysis of SQUID operation, let us supply the device with some bias ac current I_w, as in Fig. 4.21. If $\Phi_T = 0$, the output voltage will be V_{01}; if $\Phi_T = \Phi_0/2$, the output voltage will be V_{02}. The question is: how will the voltage V_0 change if we increase the input flux Φ_T from 0 to $\Phi_0/2$? Obviously, it will gradually decrease from V_{01} to V_{02}. However, if $\Phi_T = \Phi_0$, the situation is analogous to the case $\Phi_T = 0$, as one can see from Fig. 4.20, and the output will again be V_{01}. Thus we have arrived at the so-called 'triangular' dependence of the output voltage V_0 on the input flux (Fig. 4.22). It follows from this figure that the device is sensitive to variations of the magnetic flux Φ_T, with sensitivity much better than one flux quantum. Theoretical analysis of the sensitivity of rf SQUIDs has demonstrated that devices with sensitivity of the order of $10^{-5}\,\Phi_0$ are feasible, provided the measuring time is not less than 1 s.

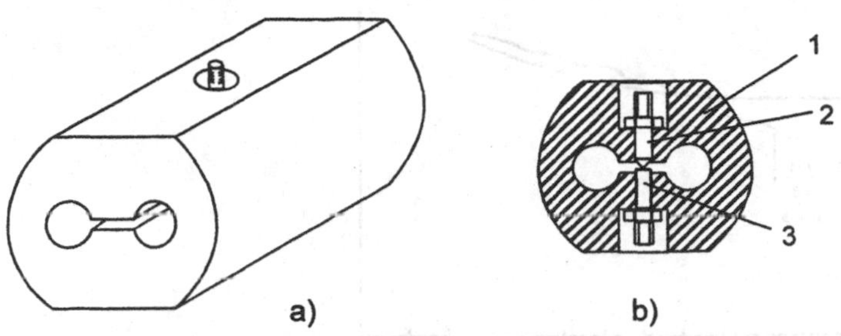

Fig. 4.23. Single-contact SQUID proposed by Zimmerman: **(a)** general view; **(b)** cross-section along the central plane: *1* – solid niobium frame, *2* – pointed niobium screw, *3* – flat niobium screw

A classic bulk SQUID design which has proved to be very useful in practice is the symmetric single-weak-link device proposed by Zimmerman (Fig. 4.23). Here the Josephson junction is implemented as a point contact. The input coil T and the coil of the resonant circuit K (see Fig. 4.19) are inserted in the two holes of the device. Nowadays, the best low- and high-T_c SQUIDs are made using the planar multilayer thin film technology [42].

4.6 Applications of Weak Superconductivity

The phenomenon of weak superconductivity – renowned for its beautiful physics – is also the basis of a large number of applications. Some of them are outlined below.

We already know that, if the current through a Josephson junction exceeds a critical value, the voltage across the junction has an ac component. The frequency of this component is defined by (4.21). It is easy to see that, if the junction is exposed to an external rf field and the frequency of the field coincides with the frequency of the Josephson radiation, some sort of resonance must appear. Indeed, in this case, the current–voltage characteristic showing the average values of the current and voltage has the form of a staircase, as in Fig. 4.24. The spacing of the constant voltage steps is exactly $\hbar\omega/2e$. Historically, these steps are called Shapiro steps [30]. Since the frequency can be measured very accurately and the quantities $\hbar$ and e are universal constants, the spacing of the voltage steps on the current–voltage characteristic can also be measured with very high accuracy. This is why the

Josephson junction can be used as a voltage standard [44]. Conversely, the ac Josephson effect can be used to build high-frequency oscillators. Due to the limited radiation power of a single Josephson junction, various types of junction arrays have been proposed in order to increase the output power and to achieve coherent emission of radiation [45].

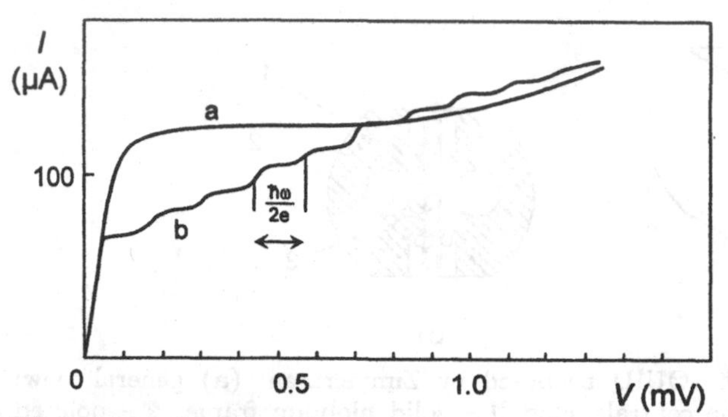

Fig. 4.24. Current–voltage characteristic of a Josephson junction: (a) without external rf electromagnetic field; (b) in the presence of an external rf field. The finite slope of the superconducting part is due to a normal film connected in series with the junction [43]

The most celebrated use of SQUIDs is as high-sensitivity magnetometers which are able to register increments of magnetic field of the order of 10^{-10} G in 1 Hz bandwidth. For comparison, recall that the Earth's magnetic field is about 0.5 G. Obviously, the availability of such sensitive devices opens new horizons in various fields of science: geology, geophysics, biophysics, and others. An example is the use of SQUIDs in pioneering studies of the magnetic fields associated with the human heart, the so-called magneto-cardiograms. It turned out that they provide a lot more information about heart problems than the usual electro-cardiograms and therefore promise to become a tool for better diagnostics in this field.

If a SQUID is used as a zero-current device incorporated in an ordinary bridge circuit, it behaves as a voltmeter with sensitivity of about 10^{-15} V. Such voltmeters are already being used in physics laboratories. In addition, Josephson junctions are used in various high-frequency detectors and sensors, in particular, as sensitive elements for detection of millimeter and submillimeter electromagnetic waves in radiotelescopes.

Finally, it is important to mention the use of Josephson junctions as logic and memory elements in digital electronics. They allow much higher operation frequencies (up to hundreds of GHz) than conventional semiconductor-based elements. Another key factor in favor of superconducting digital electronics is the reduced energy consumption as compared to semiconductors. The most promising type of logic is the so-called RSFQ (rapid single flux quantum) logic which is presently under development in many research laboratories [46].

The sphere of SQUID applications is bound to widen rapidly in the near future and at this particular moment it is impossible to foretell what wonderful results may follow.

Problems

Problem 4.1. Two Josephson junctions with critical currents $I_{c1} = 5 \times 10^{-4}$ A and $I_{c2} = 7 \times 10^{-4}$ A are connected in parallel. The total current through both junctions is 1×10^{-3} A. Find the current through each junction.

Problem 4.2. Find the difference between the maximum and minimum values of V in the plot of Fig. 4.6.

Problem 4.3. Consider a point contact having the critical current $I_c = 10^{-3}$ A and the normal state resistance $R = 2\,\Omega$. Find the value of the constant voltage across the junction, $\overline{V}$, and the frequency of Josephson radiation, ν, if the current through the contact is $I = 1.2 \times 10^{-3}$ A.

Problem 4.4. Consider a Josephson junction to which a dc current $I_0 = 7 \times 10^{-5}$ A and an ac current of amplitude $I_1 = 2 \times 10^{-6}$ A and frequency $\nu = 10$ MHz are applied. That is, the total current applied to the junction is $I = I_0 + I_1 \sin 2\pi\nu t$. The critical current of the junction is $I_c = 10^{-4}$ A. Find the voltage across the junction (in SI units).

Problem 4.5. Consider a tunnel Josephson junction formed by two Pb films. Find the critical field for penetration of a vortex into the junction, H_{c1}, and the field at the center of the vortex. The London magnetic penetration depth for Pb is $\lambda = 400$ Å and the critical current density of the junction is $j_c = 10\ \mathrm{A\,cm^{-2}}$.

Problem 4.6. Assume that the length of the tunnel junction in Problem 4.5 is $L = 0.2$ mm. Find the external fields corresponding to the first two maxima of the critical current.

Problem 4.7. Consider a single-junction interferometer of inductance 2×10^{-9} H. If the critical current of the junction decreases, the hysteresis in the dependence $\Phi(\Phi_e)$ will disappear at a certain value of the critical current. Find this value.

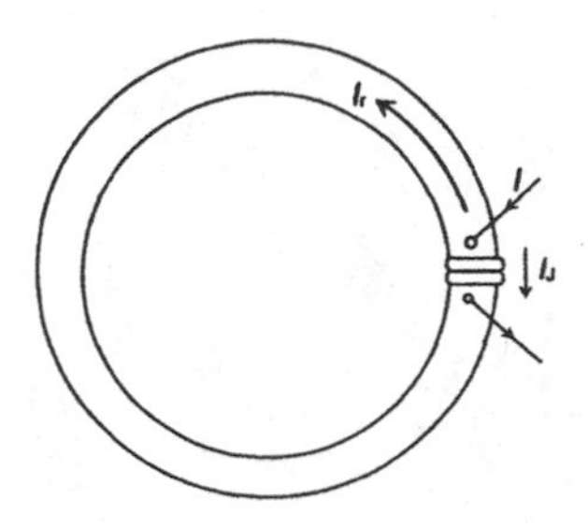

Fig. 4.25. See Problem 4.8. An external source supplies a current I, a part of which, I_r, passes through the ring and the other part, I_J, through the Josephson junction

Problem 4.8. Consider an SNS Josephson junction incorporated in a superconducting ring of inductance $L = 10^{-9}$ H. An external source supplies a total current

I, a part of which (I_r) passes through the ring and the other part (I_J) through the Josephson junction, as illustrated in Fig. 4.25. The critical current of the junction is $I_c = 10^{-6}$ A. Find the total current I.

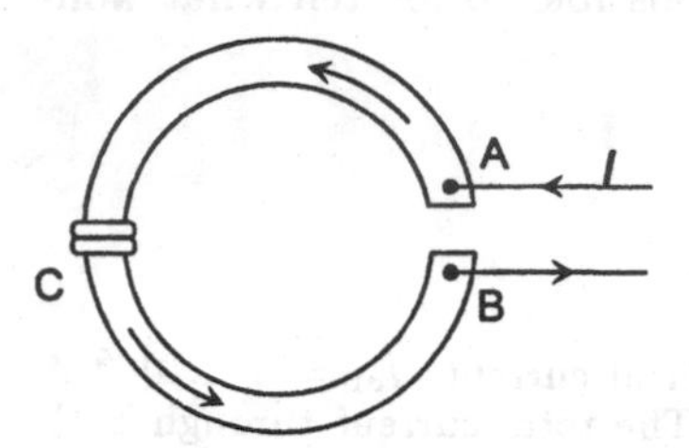

Fig. 4.26. See problem 4.9

Problem 4.9. Consider an SNS Josephson junction with the critical current $I_c = 10^{-6}$ A incorporated in a superconducting ring of inductance $L = 10^{-9}$ H. From the side opposite to the junction, a small piece of the ring is cut out, as shown in Fig. 4.26. An external current source connected to the cut supplies a current $I = 0.4 \times 10^{-6}$ A. Find the difference in the phase of the wavefunction between the points A and B, φ_{AB}. Make a plot of the current as a function of the phase difference φ_{AB}.

5. Type-II Superconductors

5.1 Introduction

The term 'type-II superconductors' was first introduced by Abrikosov in his classic paper [47] where he proposed a detailed phenomenological theory of these materials based on the GL theory and capable of explaining their magnetic properties. Later, with further development of the physics of superconductors, the theory received copious experimental verification.

Here we come across a rare case in which the development of an extensive branch of the physics of superconductors was initiated by a single theoretical work. Originally, Abrikosov's theory was greeted with a certain skepticism: so much out of the ordinary were its predictions. It was generally accepted only several years later, when it proved to be capable of explaining the complex behavior of superconducting alloys and compounds in a consistent manner. In particular, it explained the very high critical fields of some materials.

We already know that, for type-II superconductors, the energy of an interface between a normal and a superconducting region is $\sigma_{ns} < 0$. This implies that, under certain circumstances, it is energetically favorable for these materials, when placed in an external magnetic field, to become subdivided into alternating normal and superconducting domains.

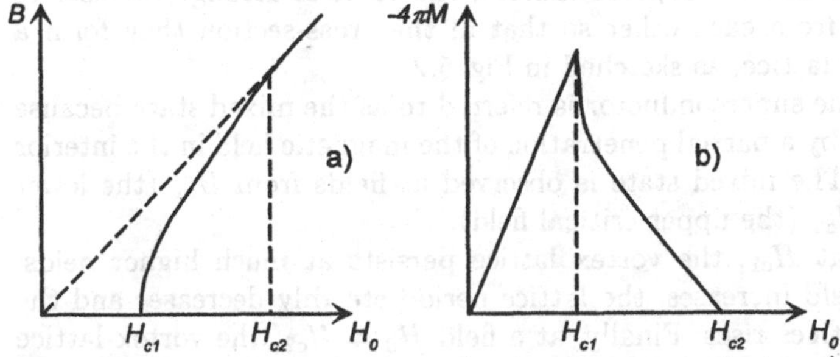

Fig. 5.1. Magnetization curves of a type-II superconductor: **(a)** magnetic induction B as a function of an external field H_0; **(b)** magnetic moment per unit volume, M, as a function of H_0

The magnetization curve of a type-II superconductor in the form of a long cylinder placed in a parallel magnetic field is shown schematically in Fig. 5.1. As long as the external field is $H_0 < H_{c1}$, the average field in the interior of the specimen is $B = 0$. However, at $H_{c1} < H_0 < H_{c2}$, a steadily increasing field B penetrates the superconductor. This field always remains below the external field H_0 and the specimen's superconductivity is not destroyed. At a certain field $H_0 = H_{c2}$, the average field in the interior, B, becomes equal to H_0 and the bulk superconductivity disappears.

Thus, above H_{c1}, a type-II superconductor does not show the Meissner–Ochsenfeld effect. Magnetic field penetrates into these materials in a very special way: as quantized vortex filaments. Every filament (or vortex) has a normal core which can be approximated by a long thin cylinder with its axis parallel to the external magnetic field. Inside the cylinder, the order parameter ψ is zero. The radius of the cylinder is of the order ξ, the coherence length. The direction of the supercurrent circulating around the normal core is such that the direction of the magnetic field generated by it coincides with that of the external field and is parallel to the normal core. The vortex current circulates within an area of radius $\sim \lambda$, the penetration depth. The size of this area can be well in excess of ξ because $\lambda \gg \xi$ for type-II superconductors.

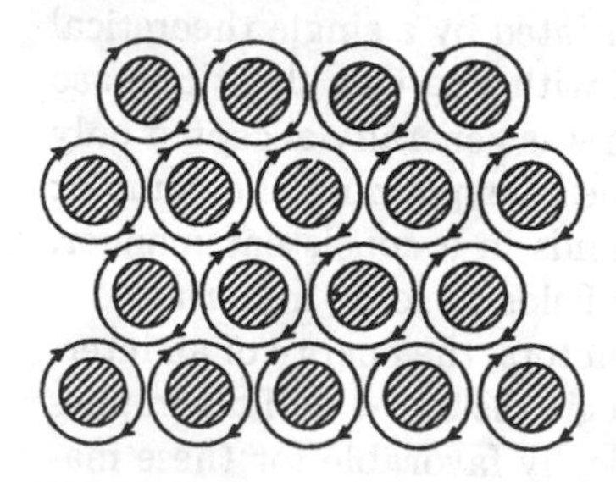

Fig. 5.2. Mixed state of a type-II superconductor. Superconducting vortices form a regular triangular lattice. Vortex cores (*dashed areas*) are normal

Each vortex carries one magnetic flux quantum. Penetration of vortices into the interior of a superconductor becomes thermodynamically favorable at $H_0 > H_{c1}$. Once inside the superconductor, the vortices arrange themselves at distances $\sim \lambda$ from each other so that in the cross-section they form a regular triangular lattice, as sketched in Fig. 5.2.

This state of the superconductor is referred to as the mixed state because it is characterized by a partial penetration of the magnetic field in the interior of the specimen. The mixed state is observed at fields from H_{c1} (the lower critical field) to H_{c2} (the upper critical field).

Once formed at H_{c1}, the vortex lattice persists at much higher fields. As the external field increases, the lattice period steadily decreases and the density of the vortices rises. Finally, at a field $H_0 = H_{c2}$, the vortex lattice grows so dense that the distance between the neighboring vortices, i.e., the lattice period, becomes of the order ξ. This means that the normal cores of the vortices come into contact with each other and the order parameter ψ

becomes zero over the entire volume of the superconductor. In other words, a second-order phase transition occurs.

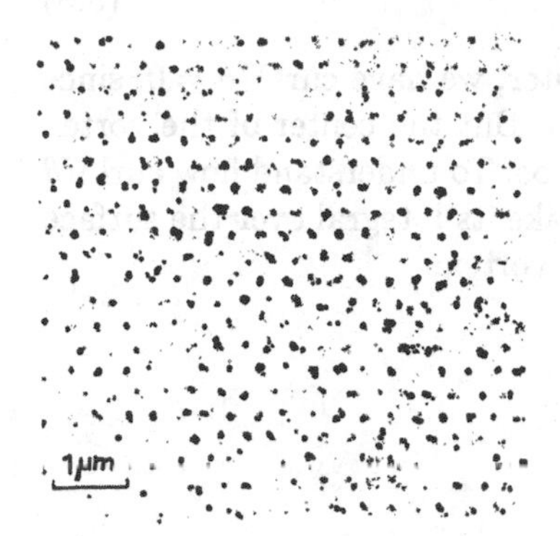

Fig. 5.3. Mixed state in niobium as observed in the electron microscope [48]

The existence of the mixed state in type-II superconductors has been reliably verified by experiment. In addition to numerous indirect proofs, a spectacular direct experimental demonstration of the vortex array is due to Träuble and Essmann [48]. They covered the top of a type-II superconducting cylinder with a thin organic film and placed it in a magnetic field so that the specimen was driven to the mixed state. Then a thin layer of fine ferromagnetic particles was sputtered onto the surface covered with the film. As the ferromagnetic particles approached the surface, they were preferentially attracted to the sites where the magnetic field lines emerged from the superconductor, i.e., to the locations of the vortices. Then, having peeled the organic film off the specimen and placed it in the electron microscope, the authors were able to enjoy a direct view of the mixed state (see Fig. 5.3).

5.2 Field of an Isolated Vortex

Let us turn to a systematic study of the mixed state. We start with the simplest case: an isolated vortex. We already know that an isolated vortex in an infinite superconductor comprises a normal core of radius $\sim \xi$ and vortex currents circulating within an area $\sim \lambda$. Assume that the GL parameter is $\kappa \gg 1$. Then $\lambda \gg \xi$. At a distance $r \gg \xi$ we have $|\psi|^2 = 1$. Let us focus on this part of the vortex. The GL equation for the vector potential (3.24) can be written as

$$\operatorname{curl}\operatorname{curl}\boldsymbol{A} = \frac{1}{\lambda^2}\left(\frac{\Phi_0}{2\pi}\nabla\theta - \boldsymbol{A}\right) . \tag{5.1}$$

Noting that $\operatorname{curl}\boldsymbol{A} = \boldsymbol{H}$, we obtain from (5.1)

$$\operatorname{curl}\boldsymbol{H} = \frac{1}{\lambda^2}\left(\frac{\Phi_0}{2\pi}\nabla\theta - \boldsymbol{A}\right) . \tag{5.2}$$

Evaluatng the curl of both parts of (5.2), we obtain

$$\boldsymbol{H} + \lambda^2 \operatorname{curl}\operatorname{curl}\boldsymbol{H} = \frac{\Phi_0}{2\pi}\operatorname{curl}\nabla\theta\,. \tag{5.3}$$

At any point of the vortex, apart from its center, we have $\operatorname{curl}\nabla\theta = 0$, since $\operatorname{curl}\nabla\varphi = 0$, where φ is an arbitrary function. But the center of the vortex represents a singularity, there we have $|\nabla\theta| \to \infty$. To understand how $\operatorname{curl}\nabla\theta$ behaves at the center of the vortex, we shall take its integral over the surface of a small circle centered on the center of the vortex:

$$\int_{\odot} \operatorname{curl}\nabla\theta \cdot \mathrm{d}\boldsymbol{S}\,.$$

By Stokes's theorem,

$$\int_{\odot} \operatorname{curl}\nabla\theta \cdot \mathrm{d}\boldsymbol{S} = \oint \nabla\theta \cdot \mathrm{d}\boldsymbol{l}\,,$$

where the path integral is taken along the circumference of our circle. Since after every full circle around the center of the vortex, the phase changes by 2π (recall that each vortex carries one flux quantum), we have

$$\int_{\odot} \operatorname{curl}\nabla\theta \cdot \mathrm{d}\boldsymbol{S} = 2\pi\,. \tag{5.4}$$

Thus the function $\operatorname{curl}\nabla\theta$ is zero everywhere except for the center of the vortex. At the center, it is infinite but its integral, according to (5.4), equals 2π. This corresponds to the behavior of the δ-function and we therefore can write

$$\operatorname{curl}\nabla\theta = 2\pi\delta(\boldsymbol{r})\boldsymbol{e}_\mathrm{v}\,,$$

where $\boldsymbol{e}_\mathrm{v}$ is the unit vector along the vortex.

As a result, (5.3) can be replaced by

$$\boldsymbol{H} + \lambda^2 \operatorname{curl}\operatorname{curl}\boldsymbol{H} = \Phi_0\,\delta(\boldsymbol{r})\,\boldsymbol{e}_\mathrm{v} \tag{5.5}$$

subject to the boundary condition $H(\infty) = 0$. The solution of (5.5) is

$$h = \frac{\Phi_0}{2\pi\lambda^2}\,K_0(r/\lambda)\,, \tag{5.6}$$

where K_0 is the MacDonald function, or the Hankel function of imaginary argument.

The asymptotic behavior of this function is:

$$K_0(z) \sim \begin{cases} \ln(1/z) & \text{at } z \ll 1\,, \\ \mathrm{e}^{-z}/z^{1/2} & \text{at } z \gg 1\,. \end{cases} \tag{5.7}$$

Thus, K_0 diverges logarithmically at small arguments and goes exponentially to zero at large ones. As follows from (5.6) and (5.7), the magnetic field at the center of the vortex is $H \to \infty$. In reality, however, this is not the case because these expressions do not hold in the vicinity of the normal core (of radius $\sim \xi$). Therefore, the field at the center of the vortex can be obtained to logarithmic accuracy by setting a cut-off at $r = \xi$:

$$H(0) \approx \frac{\Phi_0}{2\pi\lambda^2} \ln \kappa \,. \tag{5.8}$$

A more accurate value of $H(0)$ can be obtained by taking into account the actual variation of $\psi(r)$ in the vortex core and carrying out numerical integration of the GL equations [47]:

$$H(0) = \frac{\Phi_0}{2\pi\lambda^2} (\ln \kappa - 0.18) \,. \tag{5.9}$$

The above correction is not essential since all the calculations are carried out on the assumption that $\kappa \gg 1$.

The spatial variation of the magnetic field of an isolated vortex is illustrated in Fig. 5.4.

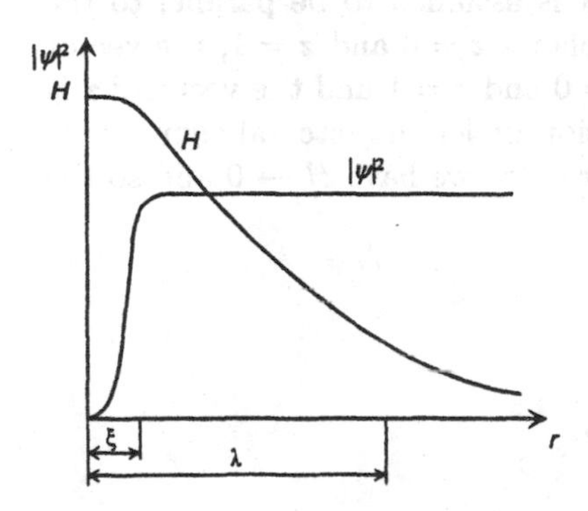

Fig. 5.4. Single vortex in an infinite superconductor. Shown are spatial distributions of the order parameter and the magnetic field of the vortex

5.3 The Lower Critical Field

Let us find the lower critical field H_{c1}, that is, the lowest field at which formation of vortices in a type-II superconductor becomes thermodynamically favorable.

First we should find the free energy of a vortex, or, more accurately, the free energy of a superconductor containing one vortex, ε, measured relative to its free energy without the vortex. We shall consider the case $\kappa \gg 1$ or $\lambda \gg \xi$. Our specimen in this case is a typical London superconductor and the corrections due to $\nabla\psi$ are negligible. Therefore, we can use the London expression for the free energy (2.8):

$$\varepsilon = \frac{1}{8\pi} \int \left[\boldsymbol{H}^2 + \lambda^2 (\operatorname{curl} \boldsymbol{H})^2 \right] \mathrm{d}V \,, \tag{5.10}$$

where the integration is carried out over the space between two infinite parallel planes perpendicular to the vortex and separated by a unit length. Then (5.10) is simply the sum of the magnetic and kinetic energies of the superconducting electrons contained in the vortex, per unit length of the vortex. Making use of the formula

$$(\mathrm{curl}\,\boldsymbol{H})^2 = \boldsymbol{H}\cdot\mathrm{curl}\,\mathrm{curl}\,\boldsymbol{H} - \mathrm{div}\,[(\mathrm{curl}\,\boldsymbol{H})\times\boldsymbol{H}]\,,$$

we have

$$\varepsilon = \frac{1}{8\pi}\int \boldsymbol{H}\cdot(\boldsymbol{H}+\lambda^2\,\mathrm{curl}\,\mathrm{curl}\,\boldsymbol{H})\,\mathrm{d}V - \lambda^2\int \mathrm{div}\,[(\mathrm{curl}\,\boldsymbol{H})\times\boldsymbol{H}]\,\mathrm{d}V\,.$$

By Gauss's theorem, we can convert the second integral into a surface integral:

$$\int \mathrm{div}\,[(\mathrm{curl}\,\boldsymbol{H})\times\boldsymbol{H}]\,\mathrm{d}V = \oint [(\mathrm{curl}\,\boldsymbol{H})\times\boldsymbol{H}]\cdot\mathrm{d}\boldsymbol{S}\,,$$

where the surface integral is taken over an infinitely remote surface and the surfaces $z=0$ and $z=1$ (the vortex is assumed to be parallel to the z axis). Since $\boldsymbol{H}$ is perpendicular to the planes $z=0$ and $z=1$, the vector $(\mathrm{curl}\,\boldsymbol{H})\times\boldsymbol{H}$ is parallel to the planes $z=0$ and $z=1$ and the vector $\mathrm{d}\boldsymbol{S}$ is perpendicular to them. Hence the expression under the integral sign is zero over these planes. On the other hand, at $r\to\infty$ we have $H\to 0$ and so the above expression is zero there, too. Then

$$\int \mathrm{div}\,[(\mathrm{curl}\,\boldsymbol{H})\times\boldsymbol{H}]\,\mathrm{d}V = 0$$

and

$$\varepsilon = \frac{1}{8\pi}\int \boldsymbol{H}\cdot(\boldsymbol{H}+\lambda^2\,\mathrm{curl}\,\mathrm{curl}\,\boldsymbol{H})\,\mathrm{d}V\,.$$

Noting that $\boldsymbol{H}$ must satisfy (5.5), we have

$$\varepsilon = \frac{\Phi_0}{8\pi}\,H(0)\,. \tag{5.11}$$

Substituting (5.8) into this expression, we obtain

$$\varepsilon = \left(\frac{\Phi_0}{4\pi\lambda}\right)^2 \ln\kappa\,. \tag{5.12}$$

In deriving (5.12), we neglected a small additional contribution to the vortex energy due to the fact that the core is normal. Indeed, the free energy density of the core exceeds that of the surrounding areas by $F_n - F_{s0} = H_{cm}^2/8\pi$, that is, by the lost condensation energy. Then, if the core radius is taken to be ξ, the additional energy disregarded in (5.12) is $(H_{cm}^2/8\pi)\,\pi\xi^2$. Using (3.38), it is easy to show that this energy is

$$\frac{1}{4}\left(\frac{\Phi_0}{4\pi\lambda}\right)^2 \ll \left(\frac{\Phi_0}{4\pi\lambda}\right)^2 \ln\kappa\,.$$

More rigorous calculations of the condensation energy lost in the core lead to the following final expression for the energy of an isolated vortex line:

$$\varepsilon = \left(\frac{\Phi_0}{4\pi\lambda}\right)^2 (\ln\kappa + 0.08)\,. \tag{5.13}$$

It follows from (5.13) that the energy of an isolated vortex is positive, that is, without an external field, the vortex cannot exist in the interior of a superconductor. Hence, if a bulk superconductor is placed in a weak external field, no vortices are formed: Their formation is not favored by the energy. Instead, the superconductor remains in the Meissner state, just as a type-I superconductor. Let us find the minimum value of the field at which vortex formation in a type-II superconductor becomes favorable.

We already know that, for a superconductor in an external magnetic field, the energy that is a minimum at equilibrium is the Gibbs free energy $\mathcal{G}$. Per unit length of the vortex, this energy is

$$\mathcal{G} = \varepsilon - \int \frac{\boldsymbol{B}\cdot\boldsymbol{H}_0}{4\pi}\,\mathrm{d}V\,, \tag{5.14}$$

where $\boldsymbol{H}_0$ is the external magnetic field and the free energy of a vortex of unit length is ε.

Indeed, by (1.22), the Gibbs free energy density is

$$G = F - \boldsymbol{B}\cdot\boldsymbol{H}_0/4\pi\,,$$

where F is the free energy density. Since $\boldsymbol{H}_0$ is applied externally, it can be taken out of the integral sign in (5.14). Then, recalling that the vortex carries one magnetic flux quantum Φ_0, we have

$$\mathcal{G} = \varepsilon - \frac{\Phi_0 H_0}{4\pi}. \tag{5.15}$$

Clearly, for a sufficiently weak external field H_0, $\mathcal{G} > 0$ and vortex formation is not favored. However, there exists such a field H_{c1}, starting from which $\mathcal{G}$ becomes negative, i.e. formation of a vortex reduces the energy. It follows from (5.15) that

$$H_{c1} = 4\pi\varepsilon/\Phi_0\,. \tag{5.16}$$

Making use of (5.13), we obtain the lower critical field for a type-II superconductor:

$$H_{c1} = \frac{\Phi_0}{4\pi\lambda^2}(\ln\kappa + 0.08)\,. \tag{5.17}$$

Comparing (5.17) and (5.8), one can see that the lower critical field is approximately half the field at the center of an isolated vortex.

Let us emphasize once again that all the above estimates are valid for superconductors with $\kappa \gg 1$ and are of logarithmic accuracy.

The field H_{c1} is a relatively weak field. For instance, if $\kappa \sim 100$ and $H_{cm} \sim 10^3$ Oe, we have $H_{c1} \sim 30$ Oe.

5.4 Interaction Between Vortices

In the preceding section we have analyzed the properties of an isolated vortex. But the number of vortices in the mixed state is usually large and they interact strongly with each other. The objective of this section is to understand how they interact.

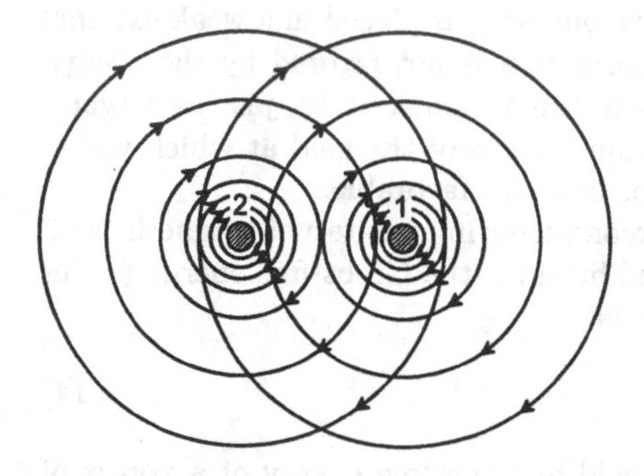

Fig. 5.5. Interaction of two parallel vortices of the same orientation

Consider two parallel vortices of the same orientation in an infinite superconductor. As before, we assume that $\kappa \gg 1$. As long as the distance between the vortices exceeds λ, they do not 'feel' each other. But when their distance becomes less than λ, the core of one vortex moves into the area where the supercurrents of the other vortex circulate, as illustrated in Fig. 5.5. It is evident from the figure that the superfluid components of the electron velocity to the right of vortex *1* and to the left of vortex *2* add up while those in between the vortices subtract from each other. This means that in the area adjoining the core of vortex *1* (to the right of it), the Bernoulli pressure is less than on its left. This results in a difference of Bernoulli pressures exerted on the core of vortex *1* in the direction from left to right. Applying similar arguments to vortex *2*, one can conclude that two parallel vortices of the same sign repel each other and the repulsion force acts only on the vortex core.

Let us now work out the force between two parallel vortices. Assume that the coordinates of the vortex cores are $\boldsymbol{r}_1$ and $\boldsymbol{r}_2$, respectively. By (2.8), the energy of a superconductor containing two vortices measured from its energy without vortices is

$$\mathcal{F} = \frac{1}{8\pi} \int \left[\boldsymbol{H}^2 + \lambda^2 (\operatorname{curl} \boldsymbol{H})^2\right] \mathrm{d}V . \tag{5.18}$$

Here $\boldsymbol{H}$ is the total magnetic field generated by the two vortices. This field must satisfy an equation constructed by analogy with (5.5):

$$\boldsymbol{H} + \lambda^2 \operatorname{curl} \operatorname{curl} \boldsymbol{H} = \Phi_0 [\delta(\boldsymbol{r} - \boldsymbol{r}_1) + \delta(\boldsymbol{r} - \boldsymbol{r}_2)]\, \boldsymbol{e}_\mathrm{v} .$$

After a series of transformations analogous to those in the foregoing section, where the energy of an isolated vortex was derived, we obtain from (5.18)

$$\mathcal{F} = \frac{\Phi_0}{8\pi}\,[H(\boldsymbol{r}_1) + H(\boldsymbol{r}_2)]\ . \tag{5.19}$$

Here $H(\boldsymbol{r}_1)$ is the field at the center of vortex *1*. It incorporates the field generated by vortex *1* itself and the field $H_{12}(x)$ due to vortex *2* (the latter is separated from vortex *1* by the distance $x = |\boldsymbol{r}_1 - \boldsymbol{r}_2|$). The same applies to the field $H(\boldsymbol{r}_2)$. Then it follows from (5.19) that

$$\mathcal{F} = 2\varepsilon + \frac{\Phi_0}{8\pi}\,2H_{12}(x)\ , \tag{5.20}$$

where ε is the energy of an isolated vortex defined by (5.13).

The physical significance of (5.20) is evident: the first term is the energy of two noninteracting vortices and the second term is the energy of their interaction. Denote the latter by $U(x)$:

$$U(x) = \frac{\Phi_0 H_{12}(x)}{4\pi}\ . \tag{5.21}$$

Then the interaction force per unit length of the vortex is

$$f = -\frac{\mathrm{d}U}{\mathrm{d}x} = -\frac{\Phi_0}{4\pi}\,\frac{\mathrm{d}H_{12}}{\mathrm{d}x}\ .$$

Taking into account that for two parallel vortices, according to Maxwell's equations,

$$\frac{\mathrm{d}H_{12}}{\mathrm{d}x} = \frac{4\pi}{c}\,j_{12}(x)\ ,$$

where $j_{12}(x)$ is the current density induced by the first vortex at the center of the second vortex (or vice versa), we have

$$|f| = \frac{1}{c}\,j_{12}\,\Phi_0\ . \tag{5.22}$$

The above formula is also valid in a more general case: If there is an externally applied current j circulating around a vortex, it exerts a Lorentz force on the vortex core, per unit length,

$$f_{\mathrm{L}} = \frac{1}{c}\,[\boldsymbol{j} \times \boldsymbol{\Phi}_0]\ , \tag{5.23}$$

where $\boldsymbol{j}$ is the density of the external current at the location of the normal core and $\boldsymbol{\Phi}_0 = \boldsymbol{e}_{\mathrm{v}}\Phi_0$.

5.5 The Upper Critical Field

The mixed state of a homogeneous type-II superconductor is characterized by a regular, usually triangular, vortex lattice. As the external field is increased, the lattice period decreases and when it becomes of the order of the coherence length ξ, a second-order phase transition occurs from the mixed to the normal state. This happens when the external field reaches the value of the upper critical field H_{c2}.

We can estimate the order of magnitude of H_{c2} even without exact calculations, proceeding from the following arguments. For two neighboring vortices at a distance ξ from each other, the distance ξ can be interpreted as the thickness of a superconducting layer separating their normal cores. It can be roughly approximated by a thin film of thickness ξ. But we know from Sect. 3.5 that a thin film in a parallel magnetic field undergoes a second-order phase transition to the normal state at the field $H_{c\|} \sim H_{cm}\lambda/d$, where d is the thickness of the film. If, as in our case, the film thickness is ξ, one should expect the transition to the normal state at an external field $\sim H_{cm}\lambda/\xi$. This provides us with a simple estimate of the upper critical field:

$$H_{c2} \sim \kappa H_{cm} . \tag{5.24}$$

Our estimate differs from the result of exact calculations only by a numerical factor $\sqrt{2}$ [47]:

$$H_{c2} = \sqrt{2}\,\kappa H_{cm} . \tag{5.25}$$

The upper critical field can be rather large. For example, for $\kappa \sim 100$ and $H_{cm} \sim 10^3$ Oe we obtain $H_{c2} \sim 10^5$ Oe.

Using (5.25), the expression $\kappa = \lambda/\xi$, and (3.38), namely, $\sqrt{2}H_{cm} = \Phi_0/2\pi\lambda\xi$, we get

$$\Phi_0 = 2\pi\xi^2 H_{c2} . \tag{5.26}$$

The last formula is very useful for determining the coherence length ξ because it relates ξ to a quantity which can be easily measured in experiment: the upper critical field.

5.6 Reversible Magnetization of a Type-II Superconductor

Let us work out an expression for the magnetic moment per unit volume of a type-II superconductor, when the latter is in the mixed state and when the external field is $H_0 \gg H_{c1}$. In other words, let us find the dependence $M(H_0)$.

This problem could be handled in the usual way by varying the energy of the system in order to find the equilibrium value of the magnetic moment. But we shall instead employ a somewhat artificial method which will allow us to shorten the computations.

Consider a bulk cylinder of a type-II superconductor, with its axis parallel to the z axis, in a longitudinal magnetic field H_0. Assume that the coherence length ξ depends only on the x coordinate and increases monotonically with x. The penetration depth λ is assumed to be independent of x. Then the critical field H_{c2} will also depend on x, namely, it will decrease with increasing x. As a result, the magnetic flux density (induction) B will become a function of x, that means that a current j will start in the interior of the sample, in the direction of the y axis:

$$j = \frac{c}{4\pi}\frac{\mathrm{d}B}{\mathrm{d}x}\,.$$

Recalling that $B = H_0 + 4\pi M$, we arrive at $M = M(x)$, that is,

$$j = c\frac{\mathrm{d}M}{\mathrm{d}x}\,.$$

Furthermore, the current exerts a Lorentz force on each vortex

$$f_{\mathrm{L}} = \frac{1}{c} j\Phi_0 = \Phi_0\frac{\mathrm{d}M}{\mathrm{d}x}\,. \tag{5.27}$$

On the other hand, since the system of vortices is at equilibrium, the force f_{L} must be compensated by another force. The origin of this second force is evident: If $\xi = \xi(x)$, the self-energy of a vortex, ε, also becomes a function of x resulting in a force $-\nabla\varepsilon$ on the vortex, where (see (5.12))

$$\varepsilon = \left(\frac{\Phi_0}{4\pi\lambda}\right)^2 \ln\kappa = \left(\frac{\Phi_0}{4\pi\lambda}\right)^2 \ln\frac{\lambda}{\xi}\,. \tag{5.28}$$

Differentiating (5.28) and equating it to the right-hand side of (5.27), we obtain a condition for equilibrium:

$$\Phi_0\frac{\mathrm{d}M}{\mathrm{d}x} = \left(\frac{\Phi_0}{4\pi\lambda}\right)^2 \frac{1}{\xi}\frac{\mathrm{d}\xi}{\mathrm{d}x}\,.$$

Following integration, we obtain

$$M = \frac{\Phi_0}{16\pi^2\lambda^2}\ln\frac{\xi}{l}\,, \tag{5.29}$$

where l is the integration constant with dimensions of length.

Now we seek the dependence $M(H_0)$. The only quantity in (5.29) depending on H_0 is l. We can find the relation between H_0 and l in the following way. Since an increase of x is followed by a monotonic increase of ξ and a monotonic decrease of H_{c2}, it is clear that, at some point x_0, the field H_{c2} becomes equal to the external field H_0. This means that at this particular point $M(x_0) = 0$, i.e., $l = \xi(x_0)$. On the other hand, the relation between ξ and H_{c2} is given by (5.26). Applying it to x_0, we have

$$2\pi l^2 H_0 = \Phi_0\,.$$

Hence the ratio ξ/l is

$$\frac{\xi}{l} = \left(\frac{H_0}{H_{c2}}\right)^{1/2}\,. \tag{5.30}$$

Substituting (5.30) in (5.29), we obtain

$$M = -\frac{\Phi_0}{16\pi^2\lambda^2} \ln \left(\frac{H_{c2}}{H_0}\right)^{1/2} ,$$

or

$$M = -\frac{\Phi_0}{32\pi^2\lambda^2} \ln \frac{H_{c2}}{H_0} . \tag{5.31}$$

From the last expression, one can derive the dependence $B(H_0)$:

$$B = H_0 - \frac{\Phi_0}{8\pi\lambda^2} \ln \frac{H_{c2}}{H_0} . \tag{5.32}$$

The above formulas are correct to logarithmic accuracy, provided $\kappa \gg 1$. Let us employ them to work out the details of the $M(H_0)$ dependence at fields H_0 close to H_{c2}. Writing H_{c2}/H_0 as

$$\frac{H_{c2} - H_0}{H_0} + 1 ,$$

we obtain, instead of (5.31),

$$M = -\frac{\Phi_0}{32\pi^2\lambda^2} \ln \left(1 + \frac{H_{c2} - H_0}{H_0}\right) \approx -\frac{\Phi_0}{32\pi^2\lambda^2} \frac{H_{c2} - H_0}{H_0} .$$

Finally, using (5.26), we have

$$-4\pi M = \frac{H_{c2} - H_0}{4\kappa^2} . \tag{5.33}$$

Thus we have found that $|M|$ decreases linearly as $H \to H_{c2}$. An exact calculation of $M(H_0)$ in this field range was carried out by Abrikosov [47] with the result

$$-4\pi M = \frac{H_{c2} - H_0}{1.16\,(2\kappa^2 - 1)} . \tag{5.34}$$

It is easy to see that at $\kappa \gg 1$ we can omit the term -1 in parenthesis. Then the exact formula (5.34) will differ from the approximate one (5.33) only by a numerical coefficient ~ 1.

Let us clarify a question which might cause some confusion: What is H_{cm} for a type-II superconductor? In a type-I superconductor, it is the thermodynamic critical field at which the superconductor goes to the normal state. Now, what happens to a type-II superconductor at H_{cm} ? The answer is: nothing special. There are no special features on the magnetization curve at H_{cm}. For a type-II superconductor, the quantity H_{cm} should be considered simply as a measure of the extent to which the superconducting state of a particular material is favored over its normal state in the absence of magnetic field:

$$F_n - F_{s0} = H_{cm}^2/8\pi .$$

Alternatively, one can say that, for a type-II superconductor, a transition to the normal state occurs when the work done to magnetize it is $F_n - F_{s0}$:

$$-\int_0^{H_{c2}} M\,\mathrm{d}H_0 = H_{cm}^2/8\pi\,.$$

From the last expression we obtain the following definition of H_{cm} for a type-II superconductor:

$$H_{cm}^2 = 2\int_0^{H_{c2}} [H_0 - B(H_0)]\,\mathrm{d}H_0\,. \tag{5.35}$$

5.7 Surface Superconductivity. The Third Critical Field

Consider a type-II superconductor in an external magnetic field H_0 decreasing from $H_0 > H_{c2}$. At the moment when H_0 falls just below H_{c2}, superconducting nuclei appear all over the volume of the superconductor and a tightly packed vortex lattice forms. The order parameter ψ at this field is small ($|\psi| \ll 1$). In other words, a phase transition of second order occurs as the material goes from the normal ($H_0 > H_{c2}$) to the mixed state ($H_0 \leq H_{c2}$).

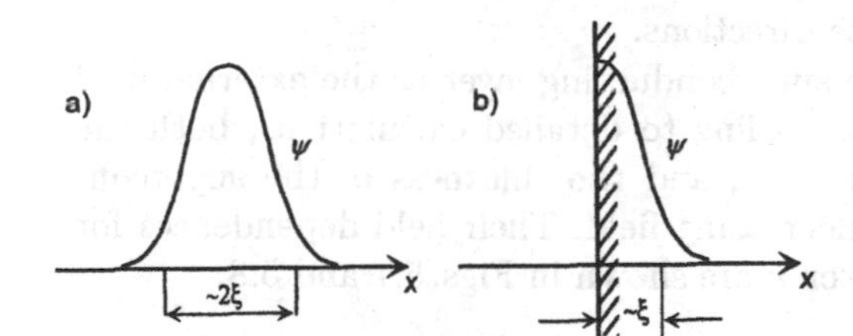

Fig. 5.6. (a) Wavefunction ψ of a nucleus of the superconducting state in an infinite superconductor, the size of the nucleus is $\sim 2\xi$; (b) the same for a superconducting semispace, the size of the nucleus is $\sim \xi$

The nucleus of superconductivity mentioned above is a region between the normal cores of two neighboring vortices. The thickness of this region is of the order $2\,\xi(T)$ because, if the vortex lattice is close-packed and the radius of the normal core is assumed to be ξ, the distance between the core centers is $\sim 2\xi$. The dependence of the order parameter on the coordinate along the line joining the vortex centers is sketched in Fig. 5.6 (a): The nucleus in this cross-section has the form of a dome-shaped curve.

This is how superconductivity arises in the bulk of a type-II superconductor at H_{c2}. It turns out, however, that at the surface of a superconductor, the superconducting state can exist in much higher fields, provided the surface is parallel to the external field.

Consider a semi-infinite superconducting space occupying the semispace $x > 0$, in an external magnetic field H_0 parallel to the z axis. Since the surface of the superconductor coincides with the plane $x = 0$, a nucleus of superconductivity, which forms at the surface, can be expected to have an effective width of the order $\xi(T)$, because $\mathrm{d}\psi/\mathrm{d}x = 0$ at the surface. Such a surface nucleus is shown in Fig. 5.6 (b). Furthermore, the critical field for the formation of such a nucleus can be expected to be larger than the critical field for nucleation of bulk superconductivity. This follows from (3.62), the expression for the critical field of a thin film, which is the most relevant reference case when considering surface superconductivity. A decrease of the film thickness results in an increase of its critical field. Therefore, we should expect the critical field for nucleation of surface superconductivity to be approximately twice the critical field for bulk superconductivity. We shall refer to this new critical field as the third critical field and denote it by H_{c3}. Numerical calculations [49] have shown that

$$H_{c3} = 1.69\, H_{c2} \,. \tag{5.36}$$

Now we can draw some conclusions. It follows from the above argumentation that, as the external magnetic field decreases and reaches the value H_{c3} from above, a thin superconducting layer appears at the surface of a superconductor. The thickness of the layer is of the order $\xi(T)$. The bulk of the superconductor remains normal and the magnetic field there is equal to the external field. In the surface superconducting layer, the magnetic field is somewhat weakened as is the case in thin superconducting films in a parallel magnetic field. The analogy with a thin film also implies the presence of screening currents circulating along both the inner and outer surfaces of the superconducting layer in opposite directions.

What happens to the surface superconducting layer as the external field decreases further, below H_{c3}? According to detailed calculations, both the amplitude of the order parameter, f_0, and the thickness of the superconducting layer, Δ, increase with decreasing field. Their field dependences for several values of the GL parameter κ are shown in Figs. 5.7 and 5.8.

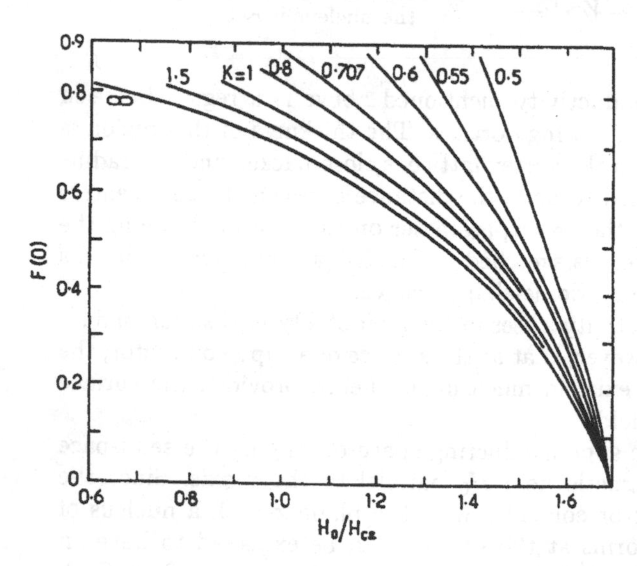

Fig. 5.7. Amplitude of the order parameter, $F(0){=}f_0$, at the surface of a superconductor versus external magnetic field, for several values of κ [50]

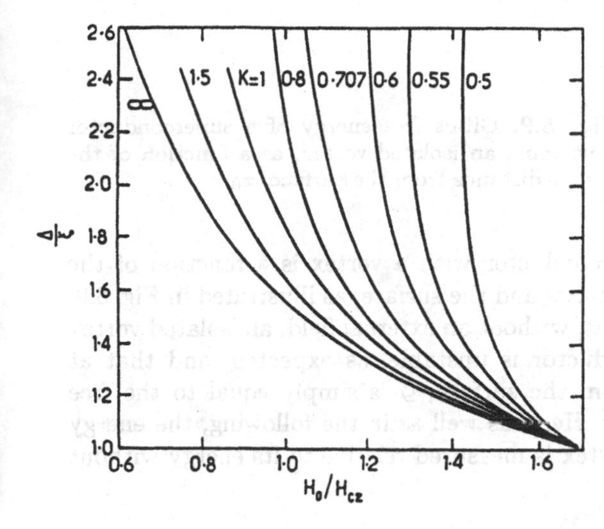

Fig. 5.8. Thickness Δ of a surface superconducting layer as a function of the external magnetic field, for several values of κ [50]

If the surface of a superconductor is covered with a layer of normal metal, it causes a reduction of H_{c3} to a value very close to H_{c2}.

It is interesting to note that the phenomenon of surface superconductivity can also be observed in some type-I superconductors. Indeed, nowhere in the above argumentation was it assumed that $\kappa > 1/\sqrt{2}$, and the result still was $H_{c3} = 1.69\,H_{c2} = 1.69\sqrt{2}\,\kappa H_{cm}$. Surface superconductivity arises if H_{c3} is larger than H_{cm}, i.e., provided

$$1.69\sqrt{2}\,\kappa H_{cm} > H_{cm}\,. \tag{5.37}$$

Thus, the condition for the existence of surface superconductivity in a type-I superconductor is

$$\kappa > \frac{1}{1.69\sqrt{2}} = 0.42\,.$$

5.8 Surface Barrier. Superheating of the Meissner State

All situations considered so far corresponded to thermodynamic equilibrium. For example, we assumed that if penetration of vortices into a type-II superconductor becomes favorable at the lower critical field, the vortices indeed penetrate at that value of the field.

However, a more detailed analysis shows that, in order to penetrate into the interior of a superconductor, vortices must first overcome an energy barrier at the surface.

Consider an ideally smooth surface of a superconductor and an isolated vortex in the interior, parallel to the surface. For convenience, we assume $\kappa \gg 1$. To begin with, the external field is assumed to be zero. The Gibbs

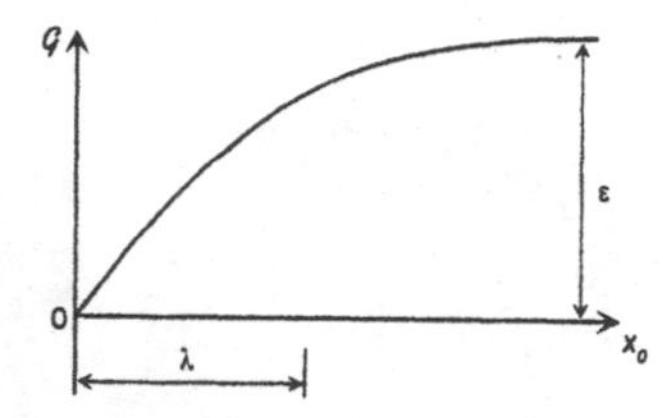

Fig. 5.9. Gibbs free energy of a superconductor containing an isolated vortex, as a function of the vortex distance from the surface, x_0

free energy $\mathcal{G}$ of such a superconductor with a vortex is a function of the distance x_0 between the vortex axis and the surface, as illustrated in Fig. 5.9. It is obvious from the figure that, without an external field, an isolated vortex in the interior of a superconductor is unstable, as expected, and that at sufficiently large distances from the surface, $\mathcal{G}$ is simply equal to the free energy of an isolated vortex, ε. Here, as well as in the following, the energy of a superconductor with a vortex is measured relative to its energy without the vortex.

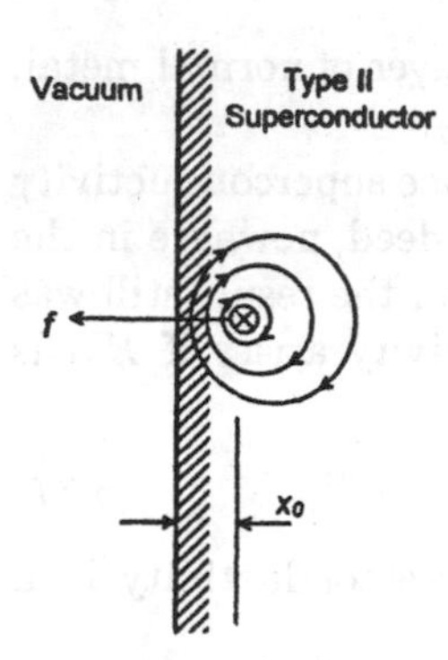

Fig. 5.10. Vortex currents near the surface of a superconductor

Let us explain the form of the function $\mathcal{G}(x_0)$. If $x_0 < \lambda$, the lines of current circulating around the vortex core are as sketched in Fig. 5.10. It is then evident that, on the left-hand side of the core, the superfluid velocity is larger than on the right-hand side, and the vortex core is subjected to the difference of Bernoulli pressures. This brings about a force pushing the vortex towards the surface. It is easy to see that the same force $f(x_0)$ can be obtained by taking the derivative

$$f = -\mathrm{d}\mathcal{G}/\mathrm{d}x_0 \,, \tag{5.38}$$

which explains the shape of $\mathcal{G}(x_0)$ in Fig. 5.9.

To find the force pushing the vortex towards the surface, we shall use the method of images. Since we have assumed $\kappa \gg 1$, the field generated by the vortex satisfies the linear equation (5.5). At the same time, at the surface, the field due to the vortex is zero (this can be proved using arguments similar to those in the well-known problem of a long solenoid: Outside the solenoid, the field is zero). Thus we have the linear equation (5.5) subject to the boundary condition of zero field at the surface. When solving a linear problem, we may apply the superposition principle and can therefore replace our problem with another, equivalent, one.

Consider an infinite superconductor containing two vortices of opposite signs at $\pm x_0$. Due to the symmetry of the problem, the field at $x = 0$ is zero, while at $x > 0$ equation (5.5) must hold. Since this differential equation has a single-valued solution, the field generated by the two vortices in the semispace $x > 0$ is exactly equal to the field which would be generated by the vortex situated at x_0 if the real surface of a superconductor were present at $x = 0$.

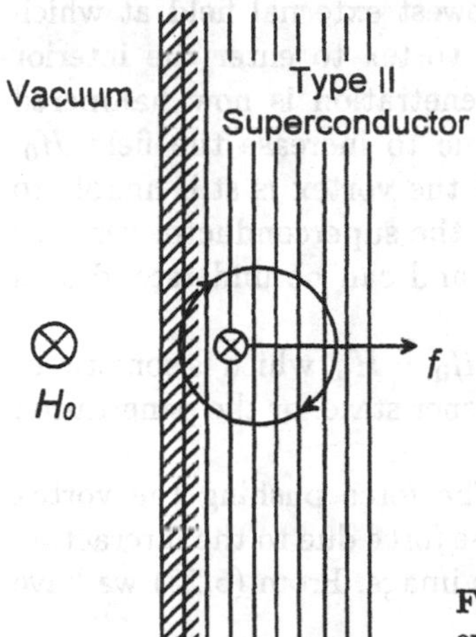

Fig. 5.11. The Meissner current, generated by an external field H_0, pushes the vortex away from the surface

Now it is easy to realize that the interaction of a vortex with the surface (attraction to the surface) can be interpreted as its interaction with its image (also attraction, because the real vortex and the image-vortex have opposite signs).

Let us now switch on an external field H_0 parallel to the surface of the superconductor, as illustrated in Fig. 5.11. This will immediately generate a Meissner current circulating near the surface which will start to push the vortex away from the surface. The result is that, on the one hand, the vortex is attracted to the surface by its own image and, on the other hand, it is repelled from it by the Meissner current. Corresponding variations of the Gibbs free energy with distance, for several values of the external field, are shown in Fig. 5.12. One can see that at $H_0 < H_{c1}$, a metastable vortex state becomes possible, such that the presence of a vortex in the superconductor is not favored thermodynamically. However, in order to leave the superconductor, the vortex must first overcome an energy barrier. This is the so-called Bean–Livingston barrier [51].

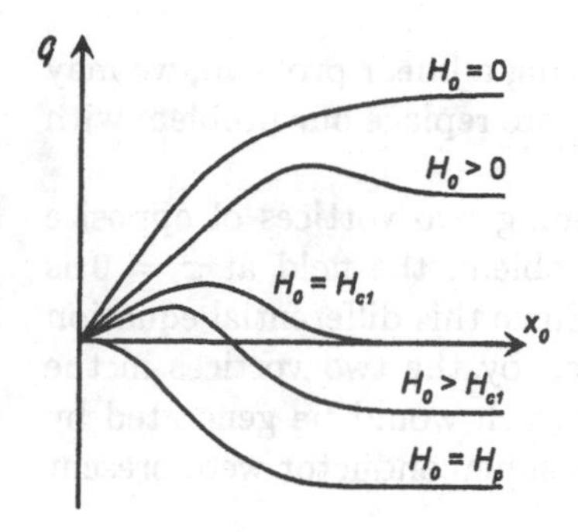

Fig. 5.12. Gibbs free energy of an isolated vortex as a function of its distance from the surface for different values of H_0

The field $H_0 = H_{c1}$ (see Fig. 5.12) is the lowest external field at which it becomes thermodynamically favorable for a vortex to enter the interior of the superconductor. However, the vortex penetration is now hampered by the Bean–Livingston barrier. As we continue to increase the field H_0, the barrier shrinks but does not disappear and the vortex is still unable to penetrate the superconductor. This means that the superconductor remains in the Meissner state which is now metastable and can be understood as a superheated Meissner state.

The barrier finally vanishes at a certain field $H_0 = H_p$, which is sometimes referred to as the superheating field for the Meissner state, or the penetration field. Let us find this field.

Assume that the vortex center is at x_0. The force pushing the vortex towards the surface is the image force, that is, the force due to the interaction of the vortex with the currents generated by its image. From (5.23) we have

$$f_{\text{image}} = \frac{1}{c}\frac{c}{4\pi}\frac{\mathrm{d}H_{\text{v}}}{\mathrm{d}x}\Phi_0 \,, \tag{5.39}$$

where H_{v} is the field generated by the image vortex and $(c/4\pi)(\mathrm{d}H_{\text{v}}/\mathrm{d}x)$ is the current that it generates at $x = x_0$. The interaction force between the vortex and the Meissner current is

$$f_{\text{M}} = \frac{1}{c}\frac{c}{4\pi}\frac{H_0}{\lambda}\,\mathrm{e}^{-x/\lambda}\Phi_0 \,, \tag{5.40}$$

where $(c/4\pi)(H_0/\lambda)\,\mathrm{e}^{-x/\lambda}$ is the current density in the Meissner state.

The Gibbs free energy is

$$\mathcal{G} = -\int f\,\mathrm{d}x \,, \tag{5.41}$$

where $f = f_{\text{M}} + f_{\text{image}}$ is the total force exerted on the vortex. Substituting (5.39) and (5.40) into (5.41) and carrying out the integration, we obtain

$$\mathcal{G}(x_0) = -\frac{\Phi_0}{4\pi}H_{\text{v}}(2x_0) + \frac{\Phi_0}{4\pi}H_0\mathrm{e}^{-x_0/\lambda} + \text{const} \,. \tag{5.42}$$

The field $H_{\text{v}}(2x_0)$ in (5.42) is the field generated at x_0 by the image vortex which is separated from this point by a distance $2x_0$. The integration constant is still to be determined. At $x_0 \to \infty$, the first two terms in (5.42) are zero. On the other hand, the energy $\mathcal{G}$ at $x_0 \to \infty$ is the Gibbs free energy of a single vortex in an infinite superconductor, which is defined by (5.15). Using (5.15) and (5.16), we obtain the integration constant in the form

$$\text{const} = \mathcal{G}(\infty) = \frac{\Phi_0}{4\pi}(H_{c1} - H_0),$$

so that

$$\mathcal{G} = \frac{\Phi_0}{4\pi}\left[H_0 e^{-x_0/\lambda} - H_v(2x_0) + H_{c1} - H_0\right]. \tag{5.43}$$

It can be easily verified that (5.43) gives a family of curves $\mathcal{G}(x_0)$ for different values of H_0 which is shown in Fig. 5.12.

The penetration field H_p can be found from a condition which is evident from Fig. 5.12:

$$\mathrm{d}\mathcal{G}/\mathrm{d}x_0{}_{|x_0=0} = 0. \tag{5.44}$$

Substituting this into the expression for $\mathcal{G}$ (5.43) and using (5.6) and (5.7), we get

$$H_p \approx H_{cm}. \tag{5.45}$$

While deriving (5.45) we had to take a derivative of $H_v(2x_0)$, which caused divergence at $x_0 = 0$. The reason is that at $x_0 = 0$ our approach is no longer appropriate because we have neglected spatial variations of the order parameter ψ. To bypass this difficulty, we assumed that the vortex emerges from the superconductor at $x_0 = \xi$, that is, when it touches the surface with its core, rather than at $x_0 = 0$. The exact calculation, due to de Gennes, leads to precisely the same result:

$$H_p = H_{cm}. \tag{5.46}$$

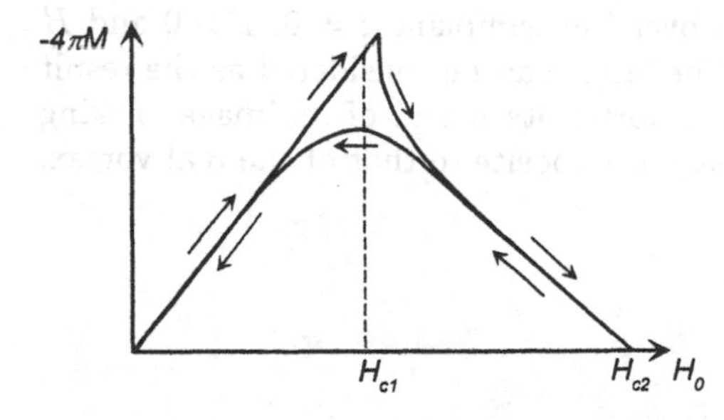

Fig. 5.13. Hysteresis in the magnetization due to the surface barrier

The existence of the Bean–Livingston barrier has been confirmed experimentally. It causes a small hysteresis in the magnetization curve of a homogeneous type-II superconductor near H_{c1}, as illustrated in Fig. 5.13. This hysteresis cannot be removed by improving bulk superconductivity. It is present even if the specimen is made ideally homogeneous.

There have also been direct measurements of H_p: de Blois and de Sorbo [52] studied Nb samples containing small amounts of oxygen, and Nb-Ta alloys. The following results were obtained for specimens with very smooth surfaces finished by electro-polishing: For oxygen-contaminated Nb, the fields were found to be $H_{c1} = 580$ Oe, $H_{c2} = 7000$ Oe, $H_{cm} = 1360$ Oe, and $H_p = 1330$ Oe; for Nb-Ta, the fields were $H_{c1} = 110$ Oe, $H_{c2} = 1600$ Oe, $H_{cm} = 310$ Oe, and $H_p = 180$–320 Oe. Comparison of the thermodynamic and penetration fields obtained in these experiments shows that (5.46) is fairly accurate.

In many cases, however, the Bean–Livingston barrier is strongly suppressed. For example, if the surface of a superconductor is rough, the vortices penetrate into it at $H_0 < H_p$.

Let us find the magnetic flux associated with a vortex parallel to the surface and positioned close to it. As follows from the general expression for the Gibbs free energy, for a superconductor with a vortex, one has

$$\mathcal{G} = \mathcal{F} - \Phi H_0/4\pi \,, \tag{5.47}$$

where $\mathcal{F}$ is the free energy of the superconductor containing one vortex (which is independent of the external field H_0) and Φ is the magnetic flux associated with the vortex. On the other hand, the Gibbs free energy $\mathcal{G}$ is given by (5.43). Taking the H_0-dependent terms in (5.43) and comparing them with (5.47), we obtain the following expression for the magnetic flux associated with the vortex near the surface:

$$\Phi = \Phi_0 \left(1 - \mathrm{e}^{-x_0/\lambda}\right) . \tag{5.48}$$

According to (5.48), as the vortex moves closer to the surface, its magnetic flux tends to zero. This result can be easily understood. The total magnetic flux of the vortex can be written as

$$\Phi = \int H \, \mathrm{d}S \,,$$

where the integration is carried out over the semiplane $z = 0$, $x > 0$ and H is the true field due to the vortex. The latter can be considered as the result of a superposition of the fields of the vortex itself and of its image. Taking into account that the sign of the image is opposite to that of the real vortex, we finally have

$$\Phi < \Phi_0 \,.$$

5.9 Some Remarks About High-Temperature Superconductors

Despite the fact that there is still no definite theory to explain their high critical temperatures, high-T_c superconductors remain superconductors in

Table 5.1. Critical temperatures and in-plane coherence lengths ξ_{ab} of some high-T_c superconductors

Material	T_c /K	ξ_{ab} /nm
$(La_{0.925}Sr_{0.075})CuO_4$	34	2.9
$YBa_2Cu_3O_7$	92.4	2.5
$Bi_2Sr_2CaCu_2O_8$	84	1.1
$Bi_2Sr_2Ca_2Cu_3O_{10}$	111	1.0
$Tl_2Sr_2Ca_2Cu_3O_{10}$	123	1.6
$HgBa_2Ca_2Cu_3O_8$	133	1.3

the classic sense. This means that the magnetic properties of these materials can be well described by the familiar BCS/Ginzburg–Landau theory. The main differences from the classic superconductors arise from intrinsic material properties such as the extremely short coherence length. Table 5.1 lists the critical temperatures and coherence lengths of some typical materials.

One of the consequences of the short coherence length was already discussed in Chap. 4. Even a grain boundary can be sufficient to suppress superconductivity, and can be used to fabricate weak-link-type Josephson devices from epitaxial films on bicrystalline substrates (see Fig. 4.1).

The second important property of high-T_c superconductors is their huge anisotropy caused by the layered crystal structure. As an example, let us consider the structure of $Bi_2Sr_2CaCu_2O_8$. The crystal is a stacked sequence

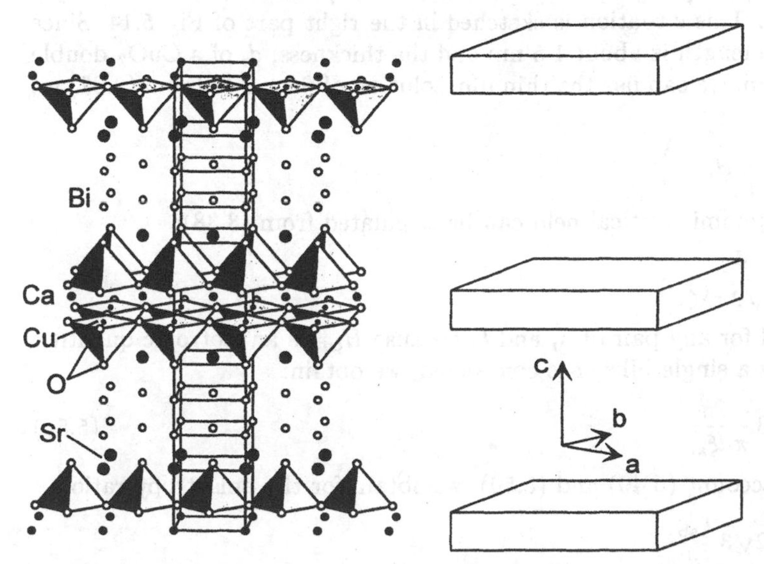

Fig. 5.14. Crystal structure of $Bi_2Sr_2CaCu_2O_8$ (left part). The Cu-O coordination polyhedra are marked for clarity. The right part shows corresponding superconducting regions

of CuO_2 planes alternating with other oxide layers (Fig. 5.14). The basic building block is the CuO_2 double layer (intercalated by Ca). These blocks are separated by two SrO and two $BiO_{1.5}$ oxide layers. In this structure, there are two principal axes, the in-plane direction and the direction perpendicular to the planes, or c axis. Consequently, we have two critical fields: $H_{c2\parallel}$ and $H_{c2\perp}$, corresponding to the field directions parallel and perpendicular to the planes. The London penetration depth and the coherence length are also anisotropic: λ_{ab}, λ_c, ξ_{ab}, ξ_c. The notation is as follows. Since, for instance, the upper critical field perpendicular to the planes, $H_{c2\perp}$, is determined by vortices whose screening currents flow parallel to the planes, we generalize the coherence length formula (5.26) to

$$H_{c2\perp} = \frac{\Phi_0}{2\pi \xi_{ab}{}^2} . \tag{5.49}$$

Correspondingly, the indices of λ and ξ, 'ab' or 'c', always indicate the direction of the screening currents.

A number of experiments indicate that high-T_c superconductors are inhomogeneous superconductors, i.e., the layered crystal structure is linked to a modulation of the order parameter [53]. Then one can argue that the superconductivity is confined to the CuO_2 planes or double planes. They are separated from neighboring planes by weakly superconducting, normal, or even insulating regions of the crystal. Three-dimensional phase coherence is provided by Josephson currents between the planes.

In order to estimate the critical field parallel to the planes, we neglect the relatively weak coupling between different double planes and consider only a single bilayer. This situation is sketched in the right part of Fig. 5.14. Since the coherence length is about 1.5 nm and the thickness, d, of a CuO_2 double plane is 0.3 nm, we can use the thin film solution of Sect. 3.5. From (3.62) we have:

$$H_{c\parallel} = 2\sqrt{6}\, H_{cm} \frac{\lambda}{d} .$$

The thermodynamic critical field can be calculated from (3.38):

$$H_{cm} = \frac{\Phi_0}{2\sqrt{2}\pi\lambda\xi} ,$$

which is valid for any pair of λ_i and ξ_i because H_{cm} is an isotropic quantity.

Since only a single bilayer is considered, we obtain:

$$H_{c\parallel} = \sqrt{3}\, \frac{\Phi_0}{\pi d \xi_{ab}} . \tag{5.50}$$

Taking into account (5.49) and (5.50), we obtain for the anisotropy ratio

$$\frac{H_{c\parallel}}{H_{c2\perp}} = 2\sqrt{3}\, \frac{\xi_{ab}}{d} .$$

This ratio is approximately 17 for $Bi_2Sr_2CaCu_2O_8$. Using the experimental value of $H_{c2\perp} = 600$ kOe, we find that the parallel critical field exceeds

10000 kOe, consistent with extrapolations from many experiments. However, in such high fields, pair-breaking effects due to spin-orbit coupling or Pauli paramagnetism are to be expected and these should decrease $H_{c\parallel}$ significantly. We note that, for a less anisotropic material, such as $YBa_2Cu_3O_7$, the assumption of independent superconducting layers might be inappropriate.

If, instead of the above argumentation, we assume a homogeneous order parameter and use the anisotropic Ginzburg–Landau description, $H_{c2\parallel}$ becomes:

$$H_{c2\parallel} = \frac{\Phi_0}{2\pi\xi_{ab}\xi_c} .$$

Then the anisotropy ratio is

$$\frac{H_{c2\parallel}}{H_{c2\perp}} = \frac{\xi_{ab}}{\xi_c} .$$

In the case of a highly anisotropic material, such as $Bi_2Sr_2CaCu_2O_8$, ξ_c would then be of the order of 0.1 nm, i.e., approaching atomic scales. In any case, this is inconsistent with the assumption of a homogeneous order parameter.

The extreme anisotropy is also responsible for many peculiar effects associated with the flux line lattice in high-T_c superconductors. A detailed analysis of these effects goes far beyond the scope of this book.

5.10 Critical Current of a Type-II Superconductor. Critical State

Consider a type-II superconductor in the mixed state to which a transport current (that is, a current from an external source) is applied in the direction perpendicular to the vortices. The current gives rise to a Lorentz force on the vortices. If the superconductor were absolutely homogeneous (free of defects), the vortices would start moving at an infinitely small Lorentz force.

Later (in Sect. 5.12) we shall find that vortex motion is accompanied by dissipation of energy. Therefore, the critical current of such an ideally homogeneous superconductor is zero.

In an inhomogeneous type-II superconductor containing various types of defects (grain boundaries, dislocation walls, dislocation tangles, voids, or second-phase precipitates), vortices can be pinned by the defects. A finite transport current is then required to set them moving, such that the Lorentz force produced by it is sufficient to tear the vortices off the defects. The latter are often referred to as pinning centers. The current density corresponding to the initiation of vortex break-off from the pinning centers is called the critical current density, j_c.

The critical current density is a structure-sensitive property and can vary by as much as several orders of magnitude as a result of thermal or mechanical treatment of the material. At the same time, the critical temperature T_c and the critical field H_{c2} can remain virtually unaffected. In superconductors which are tailored for superconducting magnets or other superconducting devices, the critical current density can be as large as $10^6\ \mathrm{A\,cm^{-2}}$.

Let us consider in more detail the flow of the critical current through a type-II superconductor containing a large number of pinning centers. In order for the transport current to be distributed over the entire cross-section of the specimen, one has to stipulate that the distribution of vortices is inhomogeneous. Indeed, the total current density at a given location is

$$\boldsymbol{j}_{\mathrm{tr}} = \frac{c}{4\pi} \operatorname{curl} \boldsymbol{B} . \tag{5.51}$$

Furthermore, $B = \Phi_0 n$ is the average field at this location, that is, the average magnetic induction in a region of dimensions much larger than the distance between the vortices; n is the average vortex density. Then it follows from (5.51) that $\boldsymbol{j}_{\mathrm{tr}}$ is nonzero only if n is a function of the coordinates: $n = n(\boldsymbol{r})$.

How does such a state, with the critical current density established across the entire cross-section of the superconductor, set in?

Assume that an infinite plate of thickness d contains a large number of defects which can pin vortices, and $d \gg \lambda$. The surfaces of the plate coincide with the planes $x = \pm d/2$. To begin with, the external magnetic field H_0 is zero and a transport current is applied in the direction of the y axis. At first, the current will flow along the surfaces of the plate (Meissner–Ochsenfeld effect), just as in the case of a homogeneous superconductor. Furthermore, as soon as the field H_I generated by this surface current exceeds the critical value H_{c1}, vortices will start to enter the plate from both surfaces, their signs being opposite when they penetrate from the opposite sides. While moving towards the center of the plate, the vortices will be pinned by the defects in the plate and, therefore, will not be able to penetrate far inside. This means that a gradient of the vortex density will appear. Evidently, it will be the maximum possible gradient, that is, the gradient corresponding to the critical value of the current. Thus, certain parts of the plate near the surfaces will carry the critical current while, in the rest of the plate, there will be no current at all. This situation is illustrated in Fig. 5.15 which shows the distribution of the magnetic field in the plate when it carries a current I_1 (per unit length along the z direction).

Let us now increase the current to a certain value I_2. The vortices, while keeping the gradient of their density at $(\nabla n)_c$, will move closer to the center so that only a small central part of the plate will remain free of transport current, while everywhere else the current density will have the critical value. If we continue to increase the current, at some moment it will reach such a value, I_c, that the current density at any point of the plate will be of the critical value, as illustrated in Fig. 5.15 (a). This is the critical state.

Let us now apply an external magnetic field H_0 parallel to the z axis, that is, parallel to the surfaces of the plate and perpendicular to the current.

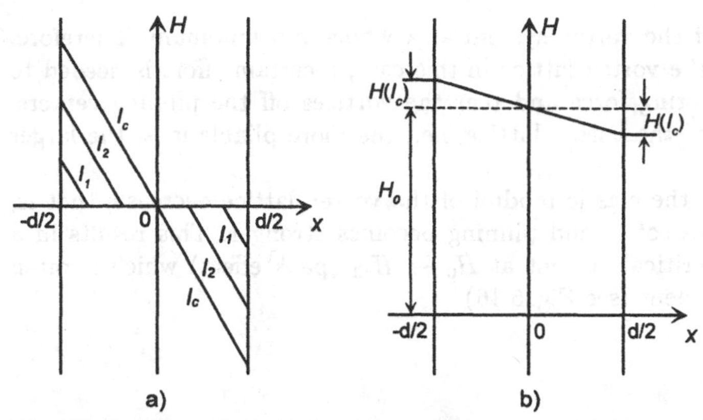

Fig. 5.15. Magnetic field distribution in a plate of thickness d containing pinning centers. The plate carries a current I: **(a)** the external field is zero and $I_1 < I_2 < I_c$ are the currents circulating in the plate; **(b)** the external magnetic field is finite. Shown is the field distribution in the critical state

Assume that the field is $H_0 \gg H_I$. Neglecting a small difference between $H_0 + H_I$ and $H_0 - H_I$, we can assume that every point of the superconductor is in the field H_0. Then the distribution of current in the critical state will be as shown in Fig. 5.15 (b). The critical gradient of the vortex density in the presence of H_0 is less than that without the field because the critical current is reduced, as explained below.

Indeed, if the Lorentz force per unit length of a vortex is $f_L = j_{tr}\Phi_0/c$, the Lorentz force per unit volume is $F_L = j_{tr}B/c$ because the density of the vortices is $n = B/\Phi_0$. When the vortices are at rest, the Lorentz force is balanced by the pinning forces exerted on the vortices. If the average density of the pinning force per unit volume is denoted by F_p, the critical current must satisfy the following equation:

$$\frac{1}{c} j_c B = F_p \,. \tag{5.52}$$

For F_p independent of the external field H_0, the critical current is $j_c \propto B^{-1}$. This dependence has indeed been observed in many experiments.

If the distribution of pinning centers in a superconductor is random, they cannot pin an absolutely rigid vortex lattice. To elucidate this statement, let us shift the lattice by a small distance in the direction of the Lorentz force. This displacement obviously does not change the energy of the system because the random distribution of pinning centers relative to the rigid vortex lattice remains essentially the same as before. And since there is no variation of energy, there is no restoring force which would push the vortices to their old positions, that is, there is no pinning.

The situation is quite different if we assume that the vortex lattice is an elastic medium. Then every time when we shift the vortex lattice, it will adjust to the given random distribution of pinning centers so as to ensure that the energy of the vortex system as a whole is a minimum. Therefore, if we try to shift the vortex lattice in this case, a certain effort is needed to overcome the restoring force and tear the vortices off the pinning centers. Clearly, the 'softer' the vortex lattice, i.e., the more pliable it is, the larger the pinning force.

At $H_0 \to H_{c2}$, the elastic moduli of the vortex lattice decrease, that is, the lattice becomes softer and pinning becomes stronger. This results in a maximum of the critical current at $H_0 \to H_{c2}$ (peak effect) which is often observed in experiment (see Fig. 5.16).

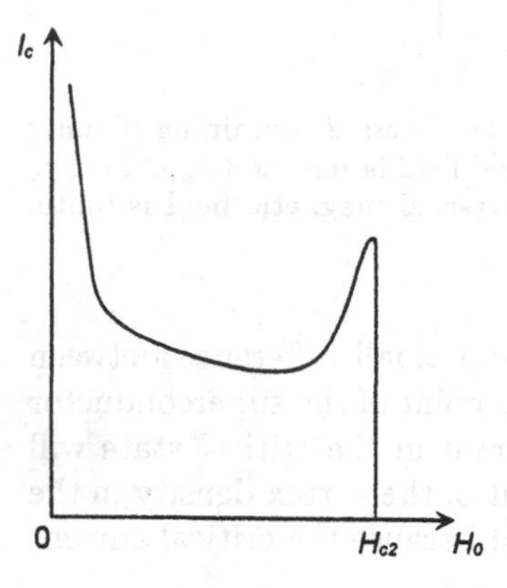

Fig. 5.16. Dependence of the critical current on perpendicular magnetic field. A pronounced peak effect is observed at $H_0 \approx H_{c2}$

5.11 Interaction of Vortices with Pinning Centers

Vortices interact with many different types of defects, which often leads to large values of the critical current. The objective of this section is to study in some detail several simple examples of defects and consider the physical mechanisms by which they interact with vortex lines. One should bear in mind that not every defect can interact with vortices effectively. In the classic superconductors, for example, vacancies, individual second-phase atoms, or other similar tiny defects are not effective as pinning centers for an obvious reason: as a rule, the characteristic size of a vortex – the coherence length – exceeds by far the atomic size, i.e., the characteristic size of such a defect. Therefore, a vortex simply 'does not notice' them. Conversely, structural defects with dimensions $\sim \xi$ and larger are very effective and can be the cause of very large critical current densities.

However, the situation is different for high-T_c superconductors. There, the coherence lengths are so small that point defects have sizes comparable to ξ. We shall discuss some aspects of this qualitatively new situation in Sect. 5.11.2.

5.11.1 Interaction of Vortices with the Surface of a Superconductor

Let us consider an ideally homogeneous type-II superconductor in the form of a plate of thickness $d \gg \lambda$. An external magnetic field H_0 is parallel to the surface of the plate; in addition, $H_{c1} \ll H_0 \ll H_{c2}$ and $\kappa \gg 1$. Then the vortices in the plate form a triangular lattice with period $a \ll d$. According to detailed calculations, the distance between the edge of the plate and the vortex row nearest to it is $\sim a$. To a good accuracy, a can be assumed constant over the plate.

As we switch on a small transport current through the plate, so that it is perpendicular to the magnetic field, the current generates a field H_I, and the total field at one side of the plate becomes $(H_0 + H_I)$ and at the other $(H_0 - H_I)$. We already know that a stable state with a gradient of the vortex density is not possible in a homogeneous superconductor. Therefore, in response to the applied current, the vortex structure must shift as a whole in the direction of the Lorentz force. Since the vortices had been at equilibrium before the current was applied, even a very small displacement from their equilibrium positions ought to bring about a restoring force. The result will be an elastic shift of the whole vortex structure by a certain distance determined by the balance between the Lorentz force and the restoring force. If the Lorentz force, acting against the restoring force, is sufficient to shift the vortices by a distance $\sim a$, a continuous flow of vortices will start. Indeed, as a vortex row comes to the edge of the plate on one side, it is annihilated by its image. At the same time, an identical row enters the plate from the opposite side and moves by a distance $\sim a$ thereby leaving enough room for a new vortex row to enter. Then a new row enters the plate, the whole structure shifts again by the distance $\sim a$, etc. This implies that the transport current shifting the vortex lattice by $\sim a$ is the critical current.

Assume that our superconductor is now in thermodynamic equilibrium. Then a magnetizing current I_M along its surface is determined by the difference between the external field H_0 at the surface and the average field B in the interior:

$$I_M = \frac{c}{4\pi}|B - H_0| = cM\ .$$

The density of the magnetizing current can be written as

$$j_M = \frac{c|M|}{\lambda}\mathrm{e}^{-x/\lambda}\ , \tag{5.53}$$

where x is the distance of a given point from the edge of the plate. A vortex occupying the position x is then acted upon by the Lorentz force

$$f_M = \frac{1}{c}j_M\Phi_0 = \frac{|M|\Phi_0}{\lambda}\mathrm{e}^{-x/\lambda}\ .$$

If the vortex shifts by a small distance δx, f_M changes by δf_M:

$$|\delta f_M| = \frac{M\Phi_0}{\lambda^2} e^{-x/\lambda} \delta x \, .$$

The force f_M is applied only to the vortices within a surface layer $\sim \lambda$. Therefore, the net change of the force on the vortex system, as it shifts by δx, is

$$\frac{|M|\,\Phi_0}{\lambda^2} \delta x \frac{\lambda}{a} \, .$$

In order to obtain the final expression for the restoring force F_{restore} per unit area of the surface, we multiply the last expression by the number of vortices within the unit length along the direction of the current, i.e., by $1/a$, and further multiply it by 2 (in order to take into account the second edge of the plate):

$$F_{\text{restore}} = \frac{2\,|M|\,\Phi_0}{\lambda a^2} \delta x \, . \tag{5.54}$$

Recalling that $a^2 \approx \Phi_0/B$, we have

$$F_{\text{restore}} = \frac{2\,|M|\,B}{\lambda} \delta x \, . \tag{5.55}$$

As was already mentioned, the condition for equilibrium is that F_{restore} equals the Lorentz force per unit area of the surface,

$$\mathcal{F}_{\text{L}} = \frac{1}{c} I_{\text{tr}} B \, , \tag{5.56}$$

where I_{tr} is the transport current per unit length of the plate. The current I_{tr} reaches its critical value when δx becomes equal to $a \approx (\Phi_0/B)^{1/2}$. Substituting this into (5.55) and equating it to (5.56) we get

$$I_{\text{c}} = \frac{2c\,|M|\,\sqrt{\Phi_0}}{\lambda\sqrt{B}} \, . \tag{5.57}$$

The average of the critical current density over the cross-section is

$$j_{\text{c}} = \frac{2c\,|M|\,\sqrt{\Phi_0}}{\lambda\sqrt{B}\,d} \, . \tag{5.58}$$

The last formula agrees well with experiment.

Let us draw some conclusions. First, we shall outline our present understanding of the flow of transport current through a superconducting plate in the mixed state. If the current is below the critical value, the Lorentz force causes an elastic shift of the vortex structure as a whole by a certain distance $\delta x < a$. As soon as this happens, the Lorentz force is balanced by a restoring force brought about by the shift, and the vortex lattice comes to rest. The lattice period is the same everywhere; therefore, in the interior of the plate, the transport current is zero. In other words, the transport current flows within the layer $\sim \lambda$ near the surface.

Let us estimate the maximum transport current, i.e., the critical current. Assume that $H_{\text{c1}} \sim 100$ Oe, $H_{\text{c2}} \sim 10^5$ Oe, $B \sim 10^4$ G, $\lambda \sim 10^{-5}$ cm, and

$d \sim 10^{-4}$ cm. Then $|M|$ is ~ 1 G. Substituting these into (5.58) we obtain

$$j_c \sim 10^5 \text{ A cm}^{-2} .$$

Thus, even a homogeneous superconducting plate in the mixed state is capable of carrying a rather large transport current. Now let us imagine a thick superconducting plate consisting of a set of thin superconducting plates, already familiar to us, which are separated from each other by thin layers of an insulator or a normal metal. In this construction, the transport current can flow along every surface of every thin plate (within the surface layer $\sim \lambda$, of course) so that the resulting transport current through our composite superconductor will be very large.

An experimental verification of (5.58) was obtained by Campbell, Evetts and Dew-Hughes [55] who studied the interaction of vortices with inclusions of a normal metal in a Pb-Bi alloy.

5.11.2 Interaction of a Vortex with a Cavity in a Superconductor

Consider an infinite superconductor containing a defect in the form of a cylindrical cavity. How will a single vortex parallel to the cavity interact with it?

Assume that the diameter of the cavity satisfies the inequality $d > \xi(T)$. Then the interaction energy can be very easily estimated. If the vortex is far away from the cavity, its normal core (of diameter $\sim 2\xi$) stores a positive energy (relative to the energy of the superconductor without the vortex) because the free energy of the normal state exceeds that of the superconducting state by $H_{\rm cm}^2/8\pi$ per unit volume. Then the energy of the normal core, per unit length, is

$$\frac{H_{\rm cm}^2}{8\pi} \pi \xi^2 . \tag{5.59}$$

On the other hand, if the vortex is trapped by the cavity, i.e., passes through its interior, it does not have a normal core and, accordingly, the energy of the system is reduced by the amount of (5.59). This means that the vortex is attracted to the cavity. The interaction force per unit length, $f_{\rm p}$, can be found easily if we recall that the energy changes by the value of (5.59) when the vortex changes its position near the edge of the cavity by $\sim \xi$:

$$f_{\rm p} \approx H_{\rm cm}^2 \xi / 8 . \tag{5.60}$$

If, instead of a cylindrical cavity, the superconductor contains a spherical cavity of diameter d, the pinning force $f_{\rm pd}$ on the vortex can be obtained from (5.60):

$$f_{\rm pd} \approx H_{\rm cm}^2 \xi d / 8 . \tag{5.61}$$

To get an idea of how large this force is, let us find the current that must be applied in the direction perpendicular to the vortex in order to produce the Lorentz force $f_{\rm L} > f_{\rm pd}$.

We know that the Lorentz force per unit length of a vortex is $j\Phi_0/c$. Then the force applied to the part of the vortex which actually interacts with the defect is $j\Phi_0 d/c$. Equating this to f_{pd} (5.61), we get

$$j = \frac{cH_{cm}^2}{8\Phi_0}\xi\,. \tag{5.62}$$

Furthermore, since $H_{cm} = \Phi_0/2\sqrt{2}\pi\lambda\xi$ (see (3.38)), we obtain from (5.62)

$$j = \frac{cH_{cm}}{16\sqrt{2}\pi\lambda}\,.$$

One can easily verify that j in the last expression is of the same order of magnitude as the pair-breaking current considered earlier in Sect. 3.6. Thus, in order to tear the vortex off a spherical void, one should apply the maximum possible current for a given superconductor.

The above reasoning is also valid in the case of a superconductor containing tiny dielectric inclusions. It will also remain valid (at least by the order of magnitude) if the inclusions are of a normal metal, provided the size of the inclusions is larger than ξ. The restriction is due to the proximity effect which is essential only at distances of the order ξ from the interface. We can then conclude that various types of normal inclusions represent effective pinning centers in superconductors. This property is widely used in technical applications when large critical currents and fields are involved.

Consider, for example, a Nb-Ti alloy used in many superconducting devices where large critical currents are of importance. A typical thermal treatment of such an alloy is as follows: a Nb-Ti wire is tempered from 800°C and then annealed for a short time ($\sim$ 30 minutes) at $\approx$ 400°C. This results in a partial transformation of the superconducting β-phase into a nonsuperconducting α-phase, in the form of tiny precipitates of α-phase in the superconducting matrix. As a result of this treatment, the critical current increases by several orders of magnitude.

Among other effective pinning centers in superconductors are dislocations, dislocation walls, grain boundaries, and interfaces between superconductors with different parameters.

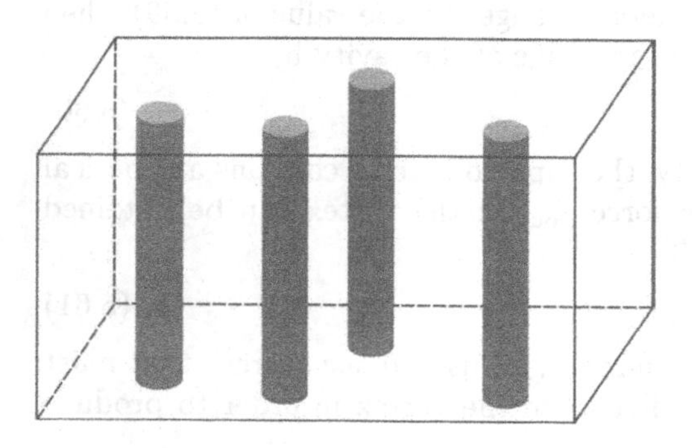

Fig. 5.17. Nuclear tracks of fast heavy ions form cylindrical amorphous regions in high-T_c superconductors

The above analysis of vortex–defect interactions forms the basis for interpreting the effects of the so-called columnar defects [56] produced in high-T_c superconductors by heavy-ion irradiation. Nuclear tracks of high-energy heavy ions form cylindrical amorphous regions in the material (see Fig. 5.17). It is generally accepted that the material in the columnar defects is not superconducting.

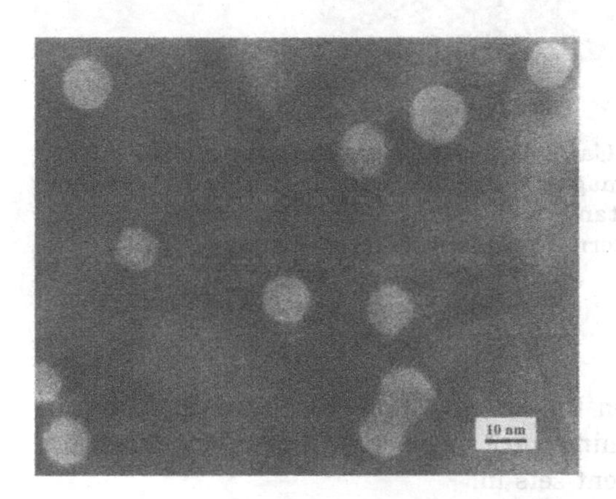

Fig. 5.18. Nuclear tracks in $Bi_2Sr_2CaCu_2O_8$ produced by irradiation with 865 MeV Pb ions. The fluence was 5×10^{10} cm^{-2}

The nuclear tracks have the useful property that their diameter is only 5–10 nm, i.e., of the order of the coherence length. As we have just seen, cavities with such sizes make extremely effective pinning centers. This is one of the very few possibilities for producing artificial defects of sizes comparable with the coherence length of high-T_c superconductors. This length is so short that conventional metallurgical methods used to produce pinning centers such as precipitates or grain boundaries are either ineffective, due to the large sizes of the defects, or not applicable at all, due to the brittleness of the materials.

Fig. 5.18 shows a transmission electron micrograph of the nuclear tracks in $Bi_2Sr_2CaCu_2O_8$ [57]. The view is along the direction of the ion beam. Columnar defects show up as bright circular spots. Each defect is able to trap approximately one vortex. Hence the optimum pinning efficiency can be expected at fields for which the vortex lattice period is less than the average distance between the amorphous tracks. Decoration patterns (c.f. Sect. 5.1) obtained on a virgin and an irradiated $Bi_2Sr_2CaCu_2O_8$ crystal (see Fig. 5.19) show the effect of columnar defects when the distance between the vortices just matches the average distance between the defects [58]. The originally regular vortex lattice is heavily deformed due to pinning by columnar defects. The defects were produced by 2.7 GeV ^{238}U ions.

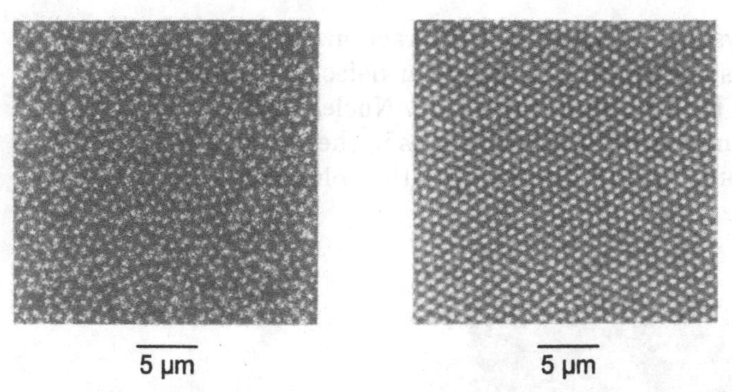

Fig. 5.19. Vortex lattice in $Bi_2Sr_2CaCu_2O_8$ with columnar defects observed by decoration (left part). The applied magnetic field of 40 Oe is such that the vortex spacing just matches the average distance between the defects. For comparison, the right part shows the decoration pattern for the unirradiated sample

5.12 Resistive State

Let us discuss what happens when the Lorentz force due to an applied current becomes larger than the pinning force and vortex flow in the direction perpendicular to the applied current sets in.

We shall show that such a vortex flow results in dissipation of energy and a finite electric resistance. This state of the vortex system is therefore referred to as the resistive state.

By Faraday's law of electromagnetic induction, a flow of magnetic flux generates an electric field E in the same direction as the applied current.[1] This means that an energy dissipation Ej_{tr} arises across the bulk of the superconductor. Assume that the velocity of the vortices in the direction of the Lorentz force, that is, in the direction perpendicular to both the current and the magnetic field, is v_{L}. Then the work done by the external source to keep the vortices moving will be $F_{\mathrm{L}}v_{\mathrm{L}}$, per unit time, where F_{L} is the density of the Lorentz force. Obviously, this work is equal to the energy dissipated by the system per unit volume and unit time, that is,

$$F_{\mathrm{L}}v_{\mathrm{L}} = Ej_{\mathrm{tr}} \,. \tag{5.63}$$

Recalling that $F_{\mathrm{L}} = j_{\mathrm{tr}}B/c$, we obtain

$$E = Bv_{\mathrm{L}}/c \,. \tag{5.64}$$

The resistivity of a superconductor resulting from the flux flow is referred to as the flux-flow resistivity and denoted by ρ_{f}:

$$\rho_{\mathrm{f}} = E/j_{\mathrm{tr}} \,. \tag{5.65}$$

The fact that the vortex motion is accompanied by dissipation of energy allows us to assume that the vortices move in a viscous medium. We introduce

[1] The Hall effect has been neglected here.

a viscosity coefficient η:

$$f_{\mathrm{friction}} = -\eta v_{\mathrm{L}} \, ,$$

where f_{friction} is the friction exerted on one vortex moving with the velocity v_{L}. Neglecting the vortex mass, we obtain an equation of motion for the vortex in the form

$$f_{\mathrm{friction}} + f_{\mathrm{L}} = 0 \, .$$

After substituting the relevant expressions for the forces per unit volume, this equation becomes

$$\frac{B}{\Phi_0} \eta v_{\mathrm{L}} = \frac{1}{c} B j_{\mathrm{tr}} \, . \tag{5.66}$$

Using (5.64) and (5.65), we can rewrite (5.66) as

$$\rho_{\mathrm{f}} = \frac{\Phi_0 B}{c^2 \eta} \, . \tag{5.67}$$

If η is independent of magnetic field, the resistivity ρ_{f} is a linear function of B. The linear dependence is indeed observed in experiments at low temperatures. An empirical formula at $T \to 0$ reads:

$$\rho_{\mathrm{n}} = \frac{\Phi_0 H_{\mathrm{c2}}}{c^2 \eta} \, .$$

It can be used to determine the viscosity coefficient at low temperatures:

$$\eta = \frac{H_{\mathrm{c2}}(0) \, \Phi_0}{c^2 \rho_{\mathrm{n}}} \, . \tag{5.68}$$

Here ρ_{n} is the normal-state resistivity of the superconductor. It is of interest to evaluate the order of magnitude of the flux flow velocity in a real experiment. Assume that the electric field in the resistive state is $E \sim 10^{-6}\ \mathrm{V\,cm^{-1}}$ and the magnetic induction is $B \sim 10^4$ G. Then, by (5.64), we have

$$v_{\mathrm{L}} = \frac{cE}{B} = 3 \times 10^{10} \, \frac{10^{-6}}{300} \, \frac{1}{10^4} = 10^{-2}\ \mathrm{cm\,s^{-1}} \, .$$

A typical current–voltage characteristic of a superconductor in the resistive state is shown in Fig. 5.20. A finite voltage appears at a certain value of current (the critical current) and, at small values of the voltage, the characteristic is nonlinear. The nonlinear part corresponds to a flow of vortices that is not yet stabilized, that is, to flux creep, where the vortices jump randomly from one pinning center to another. As the current is increased further, the current–voltage characteristic becomes linear. The linear part corresponds to the flow of the vortex structure as a whole.

There are two ways to define the critical current on an I–V characteristic. The current I_{cs} – the minimum current giving rise to a finite voltage – is the static critical current. At I_{cs}, the process of vortex break-off from pinning centers begins. The exact value of I_{cs} is difficult to determine because it

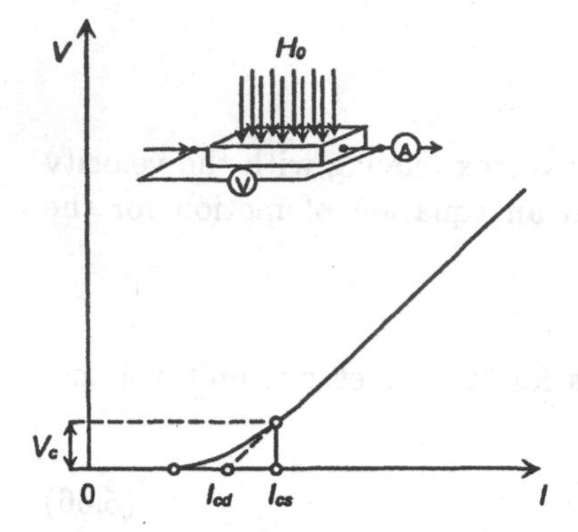

Fig. 5.20. Current–voltage characteristic of a superconductor containing crystal defects. Two ways of defining the critical current are illustrated: I_{cs} is the static and I_{cd} the dynamic critical current

depends on the actual value of the voltage that is assumed to correspond to the critical current.[2] A series of experiments [59] was intended to clarify the extent to which the critical current I_{cs} depends on the voltage threshold (or, in other words, on the sensitivity of the voltage measurements). It was found that at $H_0 \ll H_{c2}$, an enhancement of the sensitivity by several orders of magnitude leads to negligibly small changes of the critical current.

The second definition of the critical current is I_{cd}, the dynamic critical current. This value is obtained by extrapolating the linear part of a current–voltage characteristic until it intersects the I axis.

Finally, let us mention some features that are common to the resistive state of a type-II superconductor and the ac Josephson effect (Josephson radiation) [60].

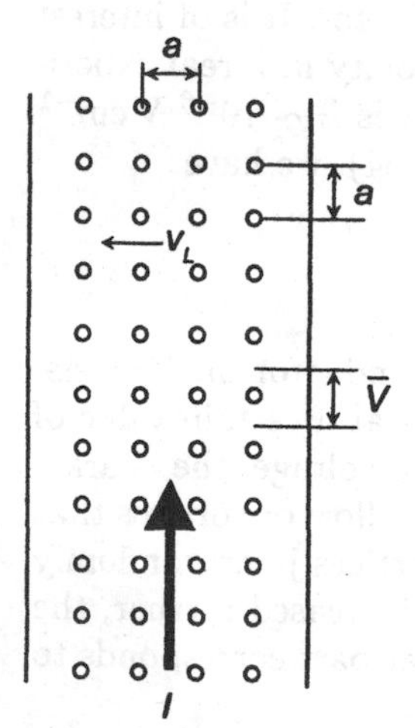

Fig. 5.21. Vortex structure in a type-II superconductor subjected to the Lorentz force moves with velocity v_L

[2] A 10^{-6} V threshold is applied in the majority of technical applications, i.e., a current that gives rise to a 10^{-6} V voltage is taken as the critical current.

Suppose that a bulk type-II superconductor is in the resistive state and the average magnetic induction in its interior is B. Assume, for the sake of simplicity, that the vortices form a square lattice as illustrated in Fig. 5.21. This assumption does not affect the generality of our discussion.

Under the Lorentz force from a current I applied to the superconductor, the vortex structure moves with velocity v_L, thereby generating an electric field

$$E = \frac{1}{c} B v_L .$$

The average voltage $\overline{V}$ over the vortex period a is then

$$\overline{V} = \frac{a}{c} B v_L \tag{5.69}$$

(see Fig. 5.21). It is evident that the vortex lattice displacement has a translational symmetry with period a. One can therefore expect that the voltage V will have an ac component of frequency

$$\omega = 2\pi \frac{v_L}{a} . \tag{5.70}$$

Substitution of this in (5.69) yields

$$\overline{V} = \frac{\omega B a^2}{2\pi c} .$$

Note that Ba^2 is the magnetic flux associated with one vortex, i.e., the flux quantum Φ_0. Therefore

$$\overline{V} = \frac{\hbar\omega}{2e} .$$

We have arrived at an expression identical to (4.21).

This radiation in type-II superconductors is very difficult to observe. Owing to the inevitable structural defects and inhmogeneities, the motion of the vortex lattice is random and turbulent, rather than coherent and laminar. However, if one takes special precautions to make the vortex motion well-ordered, it should be possible, in principle, to observe features in the resistive state of a type-II superconductor similar to those characteristic of the ac Josephson effect [61].

Problems

Problem 5.1. Two parallel superconducting vortices in an infinite superconductor are fixed at positions a and b. The distance between them is d (see Fig. 5.22). A third vortex of the same sign is allowed to move along the dashed line perpendicular to ab and passing through its middle. Find the force exerted on the third vortex as a function of the distance x from the line ab. All distances in the problem are much smaller than λ and $\kappa \gg 1$.

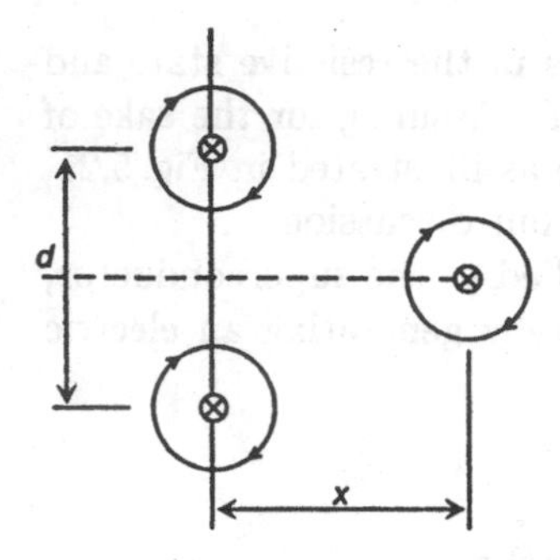

Fig. 5.22. See Problem 5.1. Two vortices are at a fixed distance d from each other. The third vortex is allowed to move along the dashed line

Problem 5.2. A superconducting alloy has an upper critical field of 30 kOe and a thermodynamic critical field of 1500 Oe. Find the magnetic field penetration depth.

Problem 5.3. Find the lower critical field of a superconducting specimen of Nb-Ta if $H_{c2} = 4000$ Oe and $\kappa = 3$.

Problem 5.4. Consider a superconductor with $\kappa \gg 1$. Find the density of vortex currents at the distance $r = \xi$ from the center of an isolated vortex.

Problem 5.5. A superconducting alloy has $H_{c2} = 150$ kOe and $\kappa = 96$. Find the energy of an isolated vortex in this alloy. Compare it with the condensation energy stored in the normal core of the vortex.

Problem 5.6. A superconducting alloy has $H_{c2} = 50$ kOe and $\kappa = 60$. Find the magnetic moment per unit volume, M, for an external field $H_0 = 10$ kOe.

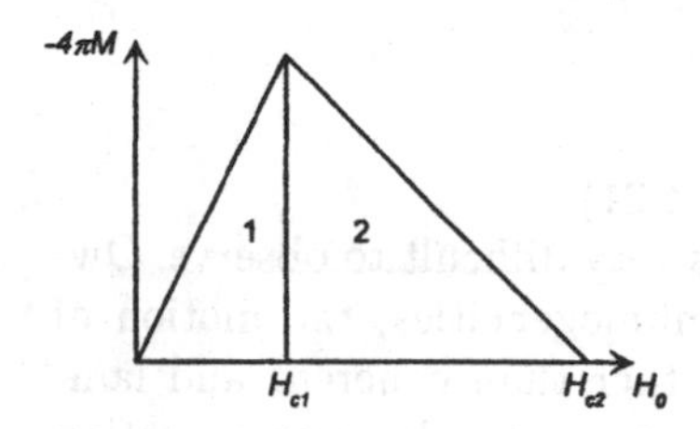

Fig. 5.23. Approximation of the magnetization curve by two triangles. Triangle (*1*) is defined by $-4\pi M = H_0$, triangle (*2*) by $-4\pi M = (H_{c2} - H_0)/[1.16\,(2\kappa^2 - 1)]$

Problem 5.7. A Nb-Ta superconducting alloy has an upper critical field $H_{c2} = 4000$ Oe and $\kappa = 3$. Approximate the magnetization curve of the alloy by two triangles, as illustrated in Fig. 5.23. Find the lower critical field H_{c1} and the thermodynamic critical field H_{cm}. Compare the value of H_{c1} with the answer to Problem 5.3 and that of H_{cm} with its true value (the latter can be found proceeding from the known value of H_{c2}).

Problem 5.8. Find the third critical field of a superconducting alloy that has $H_{cm} = 900$ Oe and $\lambda = 2500$ Å.

Problem 5.9. Find the third critical field of a superconductor with the coherence length $\xi = 90$ Å.

Problem 5.10. Find the attractive force exerted on a vortex by the surface of a flat superconductor if the vortex is parallel to the surface at a distance $l = 500$ Å, and the penetration depth is $\lambda = 3000$ Å.

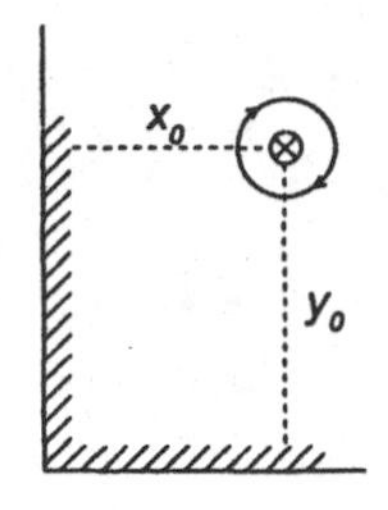

Fig. 5.24. See Problem 5.11

Problem 5.11. An isolated vortex is placed in the vicinity of the surface of a superconductor as shown in Fig. 5.24. Find the total force exerted on the vortex by the surfaces if $\lambda = 1500$ Å, $x_0 = 400$ Å, and $y_0 = 600$ Å.

Problem 5.12. Find the force with which a vortex interacts with the Meissner current if the vortex is parallel to the flat surface of a superconductor and the distance between the surface and the vortex is $l = 400$ Å. The external field is $H_0 = 50$ Oe and $\lambda = 1000$ Å.

Problem 5.13. Find the superheating field of the Meissner state if $\kappa = 24$ and $\lambda = 2000$ Å.

Problem 5.14. A perfectly homogeneous superconducting plate of thickness $l =$ 5 mm is placed in an external magnetic field $H_0 \approx B = 18$ kG. The field is parallel to the surface of the plate and perpendicular to a transport current. Find the critical current of the plate if $H_{c2} = 75$ kOe and $H_{c1} = 130$ Oe.

Problem 5.15. Estimate the flux flow resistivity ρ_f if the external field is $H_0 =$ 5×10^3 Oe, $T_c = 10$ K, and $H_{c2}(T = 4.2\,\mathrm{K}) = 40$ kOe. The remnant normal-state resistivity of the specimen is $\rho_n = 3 \times 10^{-5}\ \Omega$ cm.

Problem 5.16. A superconducting wire of length $l = 8$ cm is placed in an external field $B = 5$ T and is in the resistive state. Find the velocity v_L of moving vortices if the voltage across the wire is $U = 1.5 \times 10^{-5}$ V.

6. Microscopic Theory of Superconductivity

6.1 Introduction. Electron–Phonon Interaction

The physical mechanism of superconductivity became clear only 46 years after the phenomenon had been discovered, when Bardeen, Cooper and Schrieffer published their theory (the BCS theory) [62].

The following estimates will help us to understand how serious were the difficulties facing theoreticians at the time. The difference in free energy between the normal metal and the superconductor is, as we already know, $H_{\rm cm}^2/8\pi \sim 10^5$ erg cm^{-3}, for $H_{\rm cm} \sim 10^3$ Oe. The number of conduction electrons in 1 cm^3 is approximately 10^{22}. Then the energy responsible for superconductivity is approximately $10^5/10^{22} = 10^{-17}$ erg/electron $\sim 10^{-5}$ eV/electron. This energy should be compared with the Coulomb interaction energy ~ 1 eV which is sufficiently small to be neglected in many applications of the modern quantum theory of metals, without affecting the results. The comparison implies that it was necessary to explain the coherent behavior of electrons coupled by an interaction of energy many orders of magnitude smaller than that of other interactions, themselves often neglected exactly because of their small energies.

The first hint at the origin of superconductivity came with the discovery of the isotope effect. It was found that different isotopes of the same superconducting metal have different critical temperatures, $T_{\rm c}$, for the superconducting transition and that they obey the relation

$$T_{\rm c} M^a = \text{const} ,$$

where M is the mass of the isotope. For the majority of superconducting elements, a is close to 0.5.

Thus it became clear that the lattice of ions in a metal is an active participant in creating the superconducting state. Further theoretical analysis has demonstrated that the interaction between electrons and quantized excitations of the crystal lattice, phonons, can bring about an additional interaction between the electrons. Under certain circumstances, this interaction takes the form of electron–electron attraction. Furthermore, if the attraction proves to be stronger than the Coulomb repulsion, the electrons become effectively coupled which gives rise to the superconducting state.

Let us first try to understand how electrons can interact with each other via phonons (the latter are characterized by the energy $\hbar\omega_q$ and the wave vector $\boldsymbol{q}$).

Consider a metal at $T = 0$. Evidently, at absolute zero, there are no phonons. How is it possible then that the electrons interact via them?

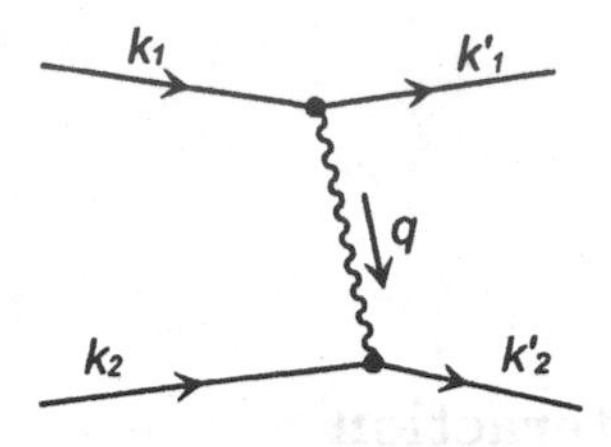

Fig. 6.1. Diagram illustrating electron–electron interaction via emission and subsequent absorption of a phonon of momentum $\hbar\boldsymbol{q}$

Consider a free electron of wave vector $\boldsymbol{k}_1$ moving in a crystal. At some moment of time it excites a lattice vibration. In other words, the electron creates a phonon (which did not exist before) and goes to a new state $\boldsymbol{k}_1'$. Let the wavevector of the created phonon be $\boldsymbol{q}$. Then, by the law of momentum conservation,

$$\boldsymbol{k}_1 = \boldsymbol{k}_1' + \boldsymbol{q} .$$

The phonon $\boldsymbol{q}$ is almost immediately absorbed by another electron $\boldsymbol{k}_2$ which goes to the state $\boldsymbol{k}_2'$ as a result. Now, what has happened? Two electrons which at first occupied the states $\boldsymbol{k}_1$ and $\boldsymbol{k}_2$, ended up in the states $\boldsymbol{k}_1'$ and $\boldsymbol{k}_2'$. This means that they were scattered by one another so that

$$\boldsymbol{k}_1 + \boldsymbol{k}_2 = \boldsymbol{k}_1' + \boldsymbol{k}_2' .$$

But two particles can be mutually scattered only if they interact with each other. Therefore, we conclude that the above process, which is illustrated in Fig. 6.1, describes an effective electron–electron interaction. Let us consider the sign of this interaction. At the moment when one of the electrons goes from the state $\boldsymbol{k}_1$ to the state $\boldsymbol{k}_1'$, it causes local oscillations of electron density at the frequency $\omega = (\bar{\varepsilon}_{\boldsymbol{k}1} - \bar{\varepsilon}_{\boldsymbol{k}1'})/\hbar$, where $\bar{\varepsilon}_{\boldsymbol{k}1}$ and $\bar{\varepsilon}_{\boldsymbol{k}1'}$ are the electron energies in the states $\boldsymbol{k}_1$ and $\boldsymbol{k}_1'$, respectively. Suppose, more specifically, that at a given moment a local electron density (negative charge) has increased. The surrounding ions will then be attracted to this location and will start moving towards it, gradually compensating the electron density increase. However, due to their large mass, the ions will continue their motion even after the density increase has been compensated and will thereby create an excess positive charge. As a result, the second electron with momentum $\boldsymbol{k}_2$ will be immediately attracted to it, that is, the particles $\boldsymbol{k}_1$ and $\boldsymbol{k}_2$ will be effectively attracted to each other. Such an attraction arises only if the lattice vibrations are in phase with the external force (its role in our case is played by the electron density oscillations of frequency $\omega = (\bar{\varepsilon}_{\boldsymbol{k}1} - \bar{\varepsilon}_{\boldsymbol{k}1'})/\hbar$). Furthermore, this can only happen if the frequency of the external force, ω, is less than the characteristic frequency of the ion system, that is, the Debye frequency (which is also the maximum possible frequency[1]). Therefore, the condition for the attraction can be formulated as $\omega < \omega_D$.

To elucidate the above statement, let us consider a simple oscillator of mass m and eigenfrequency ω_0 subjected to an external force $f\,e^{i\omega t}$. The equation of motion for such a system reads

$$\ddot{x} + \omega_0^2 x = \frac{f}{m} e^{i\omega t} . \tag{6.1}$$

We seek the solution of the form $x = x_0\, e^{i\omega t}$. Substituting it into (6.1), we have

$$x_0 = \frac{f}{\omega_0^2 - \omega^2} ,$$

that is, as long as $\omega^2 < \omega_0^2$, the oscillations $x = x_0\, e^{i\omega t}$ are in phase with the external force f. At $\omega^2 > \omega_0^2$, the oscillations are out of phase with f.

Let us go back to our electrons. To enable an electron to go from the state $\boldsymbol{k}_1$ to the state $\boldsymbol{k}_1'$, the latter must be free (due to the Pauli exclusion principle) which is possible only in the vicinity of the Fermi surface. (Recall that the latter is represented by a sphere of radius k_F in momentum space.)

Now we are ready to formulate the law of phonon-mediated interaction between electrons which forms the foundation of the BCS theory: *Electrons with energies that differ from the Fermi energy by no more than $\hbar\omega_D$ are attracted to each other.* We shall denote the energy of their interaction by $-V$. The rest of the electrons do not interact.

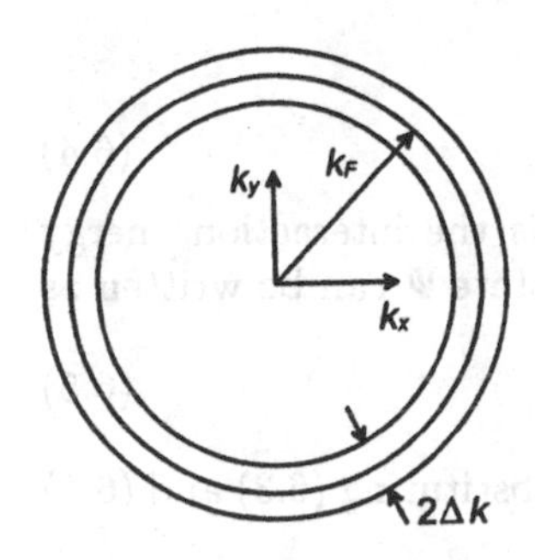

Fig. 6.2. In the BCS picture, only the electrons within the $2\Delta k$ layer near the Fermi surface interact via phonons

Let us write the matrix element of the electron interaction in the form

$$V_{\boldsymbol{k}\boldsymbol{k}'} = \begin{cases} -V, & |\bar{\varepsilon}_{\boldsymbol{k}} - \varepsilon_F| \le \hbar\omega_D, \quad |\bar{\varepsilon}_{\boldsymbol{k}'} - \varepsilon_F| \le \hbar\omega_D, \\ 0, & |\bar{\varepsilon}_{\boldsymbol{k}} - \varepsilon_F| > \hbar\omega_D, \quad |\bar{\varepsilon}_{\boldsymbol{k}'} - \varepsilon_F| > \hbar\omega_D. \end{cases} \tag{6.2}$$

[1] The existence of a maximum possible frequency for atom vibrations in crystals is easy to understand: a sound wave in a crystal is meaningless if its wavelength is less than the crystal lattice period. The existence of a minimum possible wavelength, which implies the existence of a maximum possible frequency, is then evident.

Thus, in the BCS model, only those electrons that occupy the states within a narrow spherical layer near the Fermi surface experience mutual attraction. The thickness of the layer, $2\Delta k$, is determined by the Debye energy (see Fig. 6.2):

$$\frac{\Delta k}{k_F} \sim \frac{\hbar\omega_D}{\varepsilon_F}, \qquad \varepsilon_F = \frac{\hbar^2 k_F^2}{2m}.$$

6.2 Ground State of a Superconductor

6.2.1 Electron Distribution in the Ground State

In this section, our objective is to investigate the state of a superconductor at $T = 0$, i.e., when its energy is a minimum.

First it is appropriate to recall the basic postulates of quantum mechanics which we shall need for what comes later. Suppose that $\psi_n(\boldsymbol{r}_1, \boldsymbol{r}_2, \ldots, \boldsymbol{r}_N)$ is a complete system of functions, where n runs over a set of values which we use to assign numbers to these functions. Then an arbitrary wavefunction $\Psi(\boldsymbol{r}_1, \boldsymbol{r}_2, \ldots, \boldsymbol{r}_N)$ can be expanded in a series

$$\Psi = \sum_n a_n \psi_n , \tag{6.3}$$

where a_n stands for the amplitude of the state ψ_n and $|a_n|^2$ defines the probability of finding the system in the state ψ_n.

Suppose that the Hamilton operator $\hat{H}$ is

$$\hat{H} = \hat{H}_{\text{kin}} + \hat{V} , \tag{6.4}$$

where $\hat{H}_{\text{kin}}$ is the kinetic energy operator and $\hat{V}$ is the interaction energy operator. Then the mean value of energy $\overline{E}$ in the state Ψ can be written as

$$\overline{E} = \int \Psi^* \hat{H} \Psi \, d\tau ; \tag{6.5}$$

the integration is carried out over all N variables. Substituting (6.3) and (6.4) into (6.5) yields

$$\overline{E} = \overline{E}_{\text{kin}} + \overline{V} ,$$

where

$$\overline{V} = \int \Psi^* \hat{V} \Psi \, d\tau = \int \sum_n a_n^* \psi_n^* \hat{V} \sum_m a_m \psi_m \, d\tau = \sum_{n,m} a_n^* a_m V_{nm} . \tag{6.6}$$

Here V_{nm} is the matrix element for the transition from the state ψ_m to the state ψ_n:

$$V_{nm} = \int \psi_n^* \hat{V} \psi_m \, d\tau . \tag{6.7}$$

Let us go back to the ground state of the superconductor. The ground state of the normal metal in well known. At $T = 0$, the lowest energy belongs to the state, in momentum space ($\boldsymbol{k}$ space), in which all electron states inside the Fermi surface are occupied while all those outside it are empty. The kinetic energy of such a state is obviously a minimum while the potential energy is simply absent, i.e., it is zero.

Let us now switch on the interaction energy between electrons near the Fermi level, as discussed in the foregoing section. Since this energy is associated with the effective attraction between the electrons, its contribution will be negative and the total energy of the system will be reduced. But this can happen only if the electrons are allowed to scatter from the state ($\boldsymbol{k}_1$, $\boldsymbol{k}_2$) to the state ($\boldsymbol{k}_1'$, $\boldsymbol{k}_2'$), that is, if, just before the event of scattering, the state ($\boldsymbol{k}_1$, $\boldsymbol{k}_2$) is full while the state ($\boldsymbol{k}_1'$, $\boldsymbol{k}_2'$) is empty.

Therefore, the Fermi sphere with all states below the Fermi level occupied and all states above it empty no longer corresponds to the state of minimum energy at $T = 0$.

What happens is that Nature risks a certain loss in kinetic energy in the hope of winning it back in potential energy. The total energy is now a minimum when the Fermi surface is 'smeared out', i.e., when some of the states in $\boldsymbol{k}$ space above the Fermi level are occupied while some of those below the Fermi level are empty. Moreover, when the states are being filled up, it happens in pairs: if the state $\boldsymbol{k}\uparrow$ becomes occupied, the state $-\boldsymbol{k}\downarrow$ must also be occupied (the arrows indicate the directions of electron spins). This process is illustrated in Fig. 6.3. The same applies to the empty states.

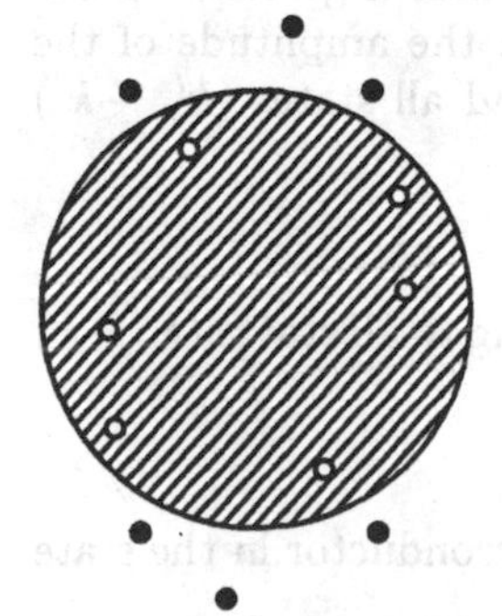

Fig. 6.3. Schematic diagram of an electron state, the energy of which can be lower than the energy of the main state (characterized by all $\boldsymbol{k}$ states below the Fermi level occupied and those above it empty)

Let us now explain why the most favorable way of coupling the electrons is such that the pairs – which are called Cooper pairs – are made of electrons with opposite momenta. A transition of an electron pair from the state ($\boldsymbol{k}_1$, $\boldsymbol{k}_2$) to the state ($\boldsymbol{k}_1'$, $\boldsymbol{k}_2'$) must obey the law of momentum conservation: $\boldsymbol{k}_1 + \boldsymbol{k}_2 = \boldsymbol{k}_1' + \boldsymbol{k}_2'$. The sum in (6.6) includes all possible transitions of this kind. The larger the number of allowed transitions, the larger the negative contribution $\overline{V}$ to the average energy of the superconductor, $\overline{E}$, will be.

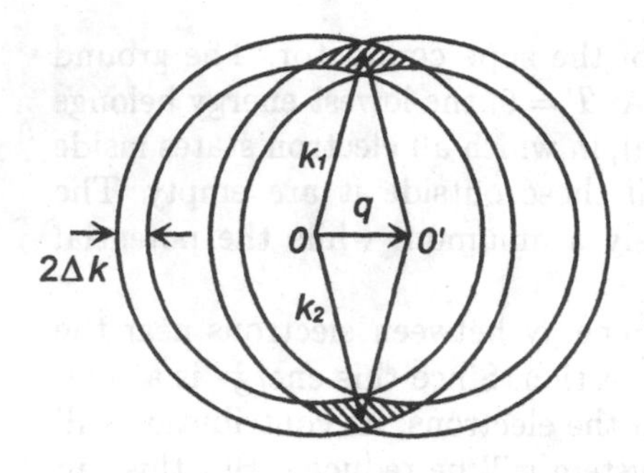

Fig. 6.4. If coupled electrons have a total momentum $\boldsymbol{q}$, the interaction involves only the electrons occupying $\boldsymbol{k}$ states within the dashed areas

Consider, for example, $\boldsymbol{k}_1 + \boldsymbol{k}_2 = \boldsymbol{q}$ as illustrated in Fig. 6.4. Here only the electrons occupying the $\boldsymbol{k}$ states in the dashed areas of momentum space are allowed to participate in the transitions. If we now reduce $\boldsymbol{q}$, the dashed areas will grow larger. Finally, at $q = 0$, all states within a band of width $\sim 2\hbar\omega_D$ near the Fermi surface will contribute to the reduction of the average energy.

Let us identify all possible states similar to those shown in Fig. 6.3. Let us assign an index n to them. Then they form a complete system of functions which can be used as a basis for a series expansion of the ground state wavefunction Ψ. The mutual scattering of two coupled electrons ($\boldsymbol{k}$, $-\boldsymbol{k}$) to the new state ($\boldsymbol{k}'$, $-\boldsymbol{k}'$) can then be described as a transition from the state ψ_n to another state ψ_m. The state ψ_n corresponds to all states ($\boldsymbol{k}$, $-\boldsymbol{k}$) full and all states ($\boldsymbol{k}'$, $-\boldsymbol{k}'$) empty, while the state ψ_m is analogous to the state ψ_n with a single exception: here all states ($\boldsymbol{k}$, $-\boldsymbol{k}$) are empty and all states ($\boldsymbol{k}'$, $-\boldsymbol{k}'$) are full (the spin indices have been omitted).

Let us introduce a new function of $\boldsymbol{k}$, $v_{\boldsymbol{k}}^2$. Suppose that it gives the probability that the pair state ($\boldsymbol{k}$, $-\boldsymbol{k}$) is occupied. Then the amplitude of the state corresponding to all states ($\boldsymbol{k}$, $-\boldsymbol{k}$) occupied and all states ($\boldsymbol{k}'$, $-\boldsymbol{k}'$) empty reads

$$a_n = \left[v_{\boldsymbol{k}}^2(1 - v_{\boldsymbol{k}'}^2)\right]^{1/2} = v_{\boldsymbol{k}} u_{\boldsymbol{k}'} , \quad \text{where} \quad u_{\boldsymbol{k}}^2 = 1 - v_{\boldsymbol{k}}^2 .$$

By analogy, the amplitude of the state ψ_m corresponding to all states ($\boldsymbol{k}$, $-\boldsymbol{k}$) empty and all states ($\boldsymbol{k}'$, $-\boldsymbol{k}'$) occupied is

$$a_m = v_{\boldsymbol{k}'} u_{\boldsymbol{k}} .$$

Using (6.6) we can write the total energy of a superconductor in the state described by the distribution $v_{\boldsymbol{k}}^2$ as

$$E_s = \sum_{\boldsymbol{k}} 2\varepsilon_{\boldsymbol{k}} v_{\boldsymbol{k}}^2 + \sum_{\boldsymbol{k},\boldsymbol{k}'} V_{\boldsymbol{k}\boldsymbol{k}'} v_{\boldsymbol{k}'} u_{\boldsymbol{k}} v_{\boldsymbol{k}} u_{\boldsymbol{k}'} . \tag{6.8}$$

The first term in (6.8) gives the total kinetic energy of the system, where $\varepsilon_{\boldsymbol{k}}$ is the energy of an electron in the state $\boldsymbol{k}$ measured from the Fermi level, that is,

$$\varepsilon_{\boldsymbol{k}} = \frac{\hbar^2 k^2}{2m} - \frac{\hbar^2 k_F^2}{2m} = \bar{\varepsilon}_{\boldsymbol{k}} - \varepsilon_F .$$

The second term is, according to (6.6), the mean potential energy of electron interaction, where the matrix element $V_{kk'}$ is defined by (6.2).

Now we should find the identity of the function v_k^2 such that the total energy E_s is a minimum. This requires that v_k^2 satisfies the equation

$$\frac{\partial E_s}{\partial v_k^2} = 0 .$$

Substituting here (6.8) and (6.2), we obtain

$$2\varepsilon_k - V\frac{1-2v_k^2}{v_k u_k}\sum_{k'}{}' v_{k'}u_{k'} = 0 .$$

It follows that

$$\frac{v_k u_k}{1-2v_k^2} = \frac{\Delta_0}{2\varepsilon_k} , \tag{6.9}$$

where

$$\Delta_0 = V\sum_k{}' v_k u_k . \tag{6.10}$$

The prime on the summation sign indicates that the summation is carried out only over the states k within the spherical layer near the Fermi surface, where the matrix element $V_{kk'}$ is nonzero (see (6.2)).

Expressing v_k^2 from (6.9) yields a quadratic equation for v_k^2:

$$v_k^4 - v_k^2 + \frac{\Delta_0^2}{4E_k^2} = 0 ,$$

where

$$E_k = \sqrt{\varepsilon_k^2 + \Delta_0^2} . \tag{6.11}$$

Then

$$v_k^2 = \frac{1}{2}\left(1 - \frac{\varepsilon_k}{E_k}\right) . \tag{6.12}$$

The minus sign in (6.12) stems from a general argument that, as $k \to 0$, we ought to have $v_k^2 \to 1$ while $\varepsilon_k \to -\varepsilon_F$.

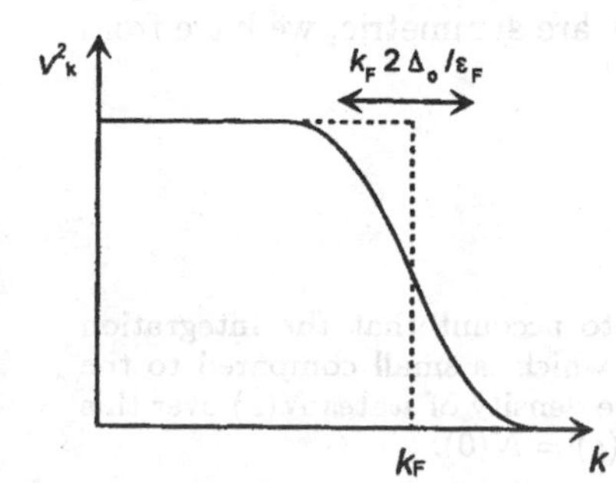

Fig. 6.5. Dependence of v_k^2 on k. At the Fermi level, $\varepsilon_k = 0$. The region where v_k^2 is 'smeared out' is $2\Delta_0$

The dependence of v_k^2 on k is plotted in Fig. 6.5. As one can see, the total energy of the system reaches its minimum when the electron distribution in the vicinity of the Fermi level is 'smeared out' over the energy interval $\sim 2\Delta_0$. Let us emphasize once more that this occurs at absolute zero! This is the identity of the ground state of the superconductor.

6.2.2 Ground-State Energy

Let us find the energy of the ground state. In order to do this, we shall first find the value of Δ_0. Substituting the expression for v_k^2, (6.12), into (6.10), we get

$$\begin{aligned}\Delta_0 &= V\sum_k{}' \left[\frac{1}{2}\left(1-\frac{\varepsilon_k}{E_k}\right)\frac{1}{2}\left(1+\frac{\varepsilon_k}{E_k}\right)\right]^{1/2} \\ &= \frac{V}{2}\sum_k{}' \left(\frac{E_k^2-\varepsilon_k^2}{E_k^2}\right)^{1/2} = \frac{V\Delta_0}{2}\sum_k{}' \left(\varepsilon_k^2+\Delta_0^2\right)^{-1/2} .\end{aligned}$$

Here we have used the definition of E_k (6.11).

Thus we obtain the equation for Δ_0:

$$1 = \frac{V}{2}\sum_k{}' \left(\varepsilon_k^2+\Delta_0^2\right)^{-1/2} . \tag{6.13}$$

Let us use the formula

$$\sum_k{}' \ldots = \int_{-\hbar\omega_D}^{\hbar\omega_D} \ldots N(\varepsilon)\,d\varepsilon$$

to go from the summation over $\boldsymbol{k}$ to an integration over ε, where $N(\varepsilon)$ is the density of states at the energy ε. Then (6.13) reduces to

$$1 = \frac{N(0)\,V}{2}\int_{-\hbar\omega_D}^{\hbar\omega_D} \left(\varepsilon^2+\Delta_0^2\right)^{-1/2} d\varepsilon . \tag{6.14}$$

Here $N(0)$ is the density of states at the Fermi level (recall that the energy ε is measured from the Fermi level[2]). Since the function under the integral sign in (6.14) is even and the integration limits are symmetric, we have from (6.14):

$$1 = N(0)\,V\int_0^{\hbar\omega_D} \left(\varepsilon^2+\Delta_0^2\right)^{-1/2} d\varepsilon .$$

[2] When going over to (6.14), we have taken into account that the integration is carried out over the energy interval ($2\hbar\omega_D$) which is small compared to the Fermi energy ε_F. Therefore, any variations of the density of states $N(\varepsilon)$ over this interval are insignificant and we can assume $N(\varepsilon) = N(0)$.

After integration we get

$$\frac{1}{N(0)\,V} = \operatorname{arcsinh}\left(\frac{\hbar\omega_{\mathrm{D}}}{\Delta_0}\right) , \tag{6.15}$$

or

$$\frac{\hbar\omega_{\mathrm{D}}}{\Delta_0} = \sinh\left(\frac{1}{N(0)\,V}\right) .$$

Recalling that for the majority of superconductors $N(0)\,V \leq 0.3$, we obtain Δ_0 in the form

$$\Delta_0 \simeq 2\hbar\omega_{\mathrm{D}} \exp\left(-\frac{1}{N(0)\,V}\right) . \tag{6.16}$$

Let us estimate Δ_0: Taking the Debye temperature $\hbar\omega_{\mathrm{D}} \sim 100$ K and $N(0)\,V \sim 0.3$, we obtain $\Delta_0 \sim 4$ K.

Now we are ready to compute the ground-state energy of a superconductor. The general expression for E_{s} is given by (6.8). In the normal state, when the interaction between the electrons is switched off and all states below the Fermi level are full, the ground-state energy is given by

$$E_{\mathrm{n}} = \sum_{k<k_{\mathrm{F}}} 2\varepsilon_{\boldsymbol{k}} . \tag{6.17}$$

Here the coefficient 2 appears because the sum is taken over pairs of states $(\boldsymbol{k}, -\boldsymbol{k})$.

The ground-state energy of the superconductor is measured from the ground-state energy of the normal metal, i.e., we seek the quantity

$$W = E_{\mathrm{s}} - E_{\mathrm{n}} .$$

Using (6.8) and (6.17), we obtain

$$W = \sum_{k<k_{\mathrm{F}}} 2\varepsilon_{\boldsymbol{k}}\,(v_{\boldsymbol{k}}^2 - 1) + \sum_{k>k_{\mathrm{F}}} 2\varepsilon_{\boldsymbol{k}} v_{\boldsymbol{k}}^2 - V {\sum_{\boldsymbol{k}\,\boldsymbol{k}'}}' v_{\boldsymbol{k}} u_{\boldsymbol{k}} v_{\boldsymbol{k}'} u_{\boldsymbol{k}'} . \tag{6.18}$$

The prime on the last summation sign indicates that the sum is taken over the layer $\pm\hbar\omega_{\mathrm{D}}$ near the Fermi surface. Substituting here (6.12) yields, after elementary transformations,

$$\begin{aligned} W &= \sum_{k<k_{\mathrm{F}}} |\varepsilon_{\boldsymbol{k}}| \left(1 - \frac{|\varepsilon_{\boldsymbol{k}}|}{E_{\boldsymbol{k}}}\right) + \sum_{k>k_{\mathrm{F}}} \varepsilon_{\boldsymbol{k}} \left(1 - \frac{\varepsilon_{\boldsymbol{k}}}{E_{\boldsymbol{k}}}\right) \\ &\quad - V {\sum_{\boldsymbol{k}\,\boldsymbol{k}'}}' v_{\boldsymbol{k}} u_{\boldsymbol{k}} v_{\boldsymbol{k}'} u_{\boldsymbol{k}'} = 2 \sum_{k>k_{\mathrm{F}}} \varepsilon_{\boldsymbol{k}} \left(1 - \frac{\varepsilon_{\boldsymbol{k}}}{E_{\boldsymbol{k}}}\right) - V {\sum_{\boldsymbol{k}\,\boldsymbol{k}'}}' v_{\boldsymbol{k}} u_{\boldsymbol{k}} v_{\boldsymbol{k}'} u_{\boldsymbol{k}'} . \end{aligned}$$

Taking into account the definition of Δ_0 (6.10), we find

$$ {\sum_{\boldsymbol{k}\,\boldsymbol{k}'}}' v_{\boldsymbol{k}} u_{\boldsymbol{k}} v_{\boldsymbol{k}'} u_{\boldsymbol{k}'} = \frac{\Delta_0^2}{V^2} .$$

Then

$$W = 2\sum_{k>k_{\mathrm{F}}} \varepsilon_k \left(1 - \frac{\varepsilon_k}{E_k}\right) - \frac{\Delta_0^2}{V}\,.$$

Going over from summation to integration, we have

$$W = 2N(0) \int_0^{\hbar\omega_{\mathrm{D}}} \varepsilon \left(1 - \frac{\varepsilon}{\sqrt{\varepsilon^2 + \Delta_0^2}}\right) \mathrm{d}\varepsilon - \frac{\Delta_0^2}{V}\,.$$

Integrating yields

$$\begin{aligned} W &= N(0)\,\Delta_0^2 \left\{ \left(\frac{\hbar\omega_{\mathrm{D}}}{\Delta_0}\right)^2 \right. \\ &\quad \left. - \frac{\hbar\omega_{\mathrm{D}}}{\Delta_0}\left[1 + \left(\frac{\hbar\omega_{\mathrm{D}}}{\Delta_0}\right)^2\right]^{1/2} + \operatorname{arcsinh}\frac{\hbar\omega_{\mathrm{D}}}{\Delta_0} \right\} - \frac{\Delta_0^2}{V}\,. \end{aligned}$$

Using (6.15), we obtain for $\hbar\omega_{\mathrm{D}} \gg \Delta_0$ the final result:

$$W = -\frac{1}{2}\,N(0)\,\Delta_0^2\,. \tag{6.19}$$

Thus, at $T = 0$, the difference in energy between the superconducting and the normal state is negative, that is, the superconducting state is more favorable energetically. And, of course, it is the state of lower energy that will be preferred by Nature.

At the beginning of this book we established that the difference in free energy between the normal and the superconducting state equals $H_{\mathrm{cm}}^2/8\pi$, where H_{cm} is the thermodynamic critical field. It then follows that at $T = 0$

$$\frac{H_{\mathrm{cm}}^2(0)}{8\pi} = \frac{1}{2}\,N(0)\,\Delta_0^2\,,$$

or

$$H_{\mathrm{cm}}(0) = \Delta_0\sqrt{4\pi N(0)}\,. \tag{6.20}$$

Thus we have expressed the thermodynamic critical field in terms of characteristic parameters of the superconductor, i.e., parameters describing its electron spectrum and electron–phonon interaction.

Let us check whether (6.20) gives reasonable orders of magnitude for relevant physical quantities. 1 cm^3 of metal contains $\sim 10^{22}$ electrons and the width of the conduction band is ~ 10 eV $= 10 \times 1.6 \times 10^{-12}$ erg. Then the density of states is $N(0) \sim 10^{22}/(10 \times 1.6 \times 10^{-12}) \sim 10^{33}\ \mathrm{erg}^{-1}\mathrm{cm}^{-3}$. According to our previous estimates, $\Delta_0 \sim 10$ K $\sim 10^{-15}$ erg. This leads to $H_{\mathrm{cm}} \sim 10^{-15} \times \sqrt{10^{34}} = 100$ Oe, which is a reasonable value for the thermodynamic critical field.

6.3 Elementary Excitation Spectrum of the Superconductor

6.3.1 The Energy Gap

We shall now acquaint ourselves with one of the most important concepts of the microscopic theory of superconductivity: the energy gap in the elementary excitation spectrum of the superconductor.

Let us focus on an arbitrary pair of states $(\boldsymbol{q}, -\boldsymbol{q})$ in a superconductor, in momentum space. First we find the contribution of this pair, $w_{\boldsymbol{q}}$, to the total energy of the superconductor. One can see from (6.8) that

$$w_{\boldsymbol{q}} = 2\varepsilon_{\boldsymbol{q}} v_{\boldsymbol{q}}^2 - 2V v_{\boldsymbol{q}} u_{\boldsymbol{q}} \sum_{\boldsymbol{k}}{}' v_{\boldsymbol{k}} u_{\boldsymbol{k}} \,. \tag{6.21}$$

Here the first term is the kinetic energy of the pair $(\boldsymbol{q}, -\boldsymbol{q})$ and the second term is the pair's contribution to the negative part of the ground-state energy. The latter arises from various interactions in which our pair participates. As a result of these interactions, the pair can go to any other state $(\boldsymbol{k}, -\boldsymbol{k})$ or, conversely, any other pair $(\boldsymbol{k}, -\boldsymbol{k})$ can go to the state $(\boldsymbol{q}, -\boldsymbol{q})$ which we have earmarked for our analysis. The coefficient 2 in the second term takes into account that the pair $(\boldsymbol{q}, -\boldsymbol{q})$ is encountered twice as the summation in (6.8) is carried out: first time when the sum runs over $\boldsymbol{k}$, and the second time when it runs over $\boldsymbol{k}'$. Using the expression for Δ_0 (6.10), as well as (6.11) and (6.12), we obtain

$$\begin{aligned} w_{\boldsymbol{q}} &= 2\varepsilon_{\boldsymbol{q}} \frac{1}{2}\left(1 - \frac{\varepsilon_{\boldsymbol{q}}}{E_{\boldsymbol{q}}}\right) - 2\left[\frac{1}{4}\left(1 - \frac{\varepsilon_{\boldsymbol{q}}^2}{E_{\boldsymbol{q}}^2}\right)\right]^{1/2} \Delta_0 \\ &= \varepsilon_{\boldsymbol{q}} - \frac{\varepsilon_{\boldsymbol{q}}^2}{E_{\boldsymbol{q}}} - \frac{\Delta_0^2}{E_{\boldsymbol{q}}} = \varepsilon_{\boldsymbol{q}} - E_{\boldsymbol{q}} \,. \end{aligned} \tag{6.22}$$

Let us now make use of (6.22). Suppose that the pair state $(\boldsymbol{q}, -\boldsymbol{q})$ in the ground state of a superconductor is empty. How will the energy of the system change if we add one more electron to it from outside and place it in the state $\boldsymbol{q}$? Since we then have a single electron in the state $\boldsymbol{q}$, the pair state $(\boldsymbol{q}, -\boldsymbol{q})$ is not allowed to take part in the scattering events, that is, it cannot contribute to the ground-state energy of the superconductor. Its potential contribution is known to us, it equals $w_{\boldsymbol{q}}$. Hence the energy of the superconductor with one 'extra' electron in the state $\boldsymbol{q}$ will be

$$W_{\boldsymbol{q}} = W - w_{\boldsymbol{q}} + \varepsilon_{\boldsymbol{q}} \,. \tag{6.23}$$

We shall refer to this 'extra' uncoupled electron as an elementary excitation of our system, or quasiparticle. The quantity W in (6.23) is the ground-state energy of the superconductor and the third term accounts for the kinetic energy of our 'extra' electron. Substituting (6.22) into (6.23) yields

$$W_q = W + E_q \,. \tag{6.24}$$

This relation is very important. Since, according to (6.11),

$$E_q = \sqrt{\varepsilon_q^2 + \Delta_0^2}\,,$$

it is evident that, by adding one extra electron to a superconductor in the ground state, we increase the energy of the system by at least the value of Δ_0 (the minimum increase corresponds to $\varepsilon_q = 0$, that is, to the state q being *on* the Fermi surface). This means that the spectrum of elementary excitations of the superconductor is separated from the ground-state energy level by an energy gap.

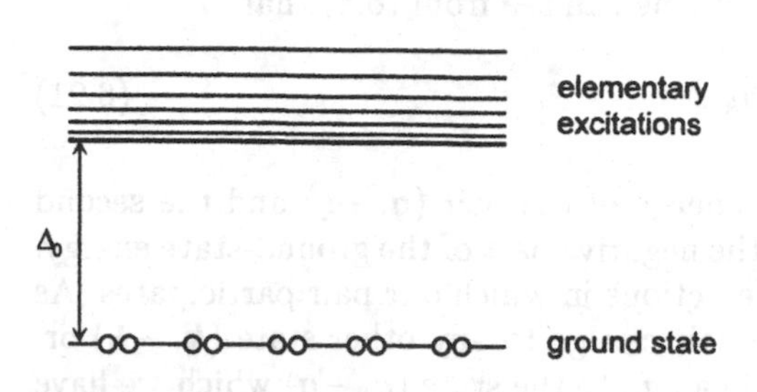

Fig. 6.6. The energy gap Δ_0 separates the energy levels of elementary excitations from the ground-state level (the energy level of electron pairs in the condensate)

To elucidate this statement, let us go back to the superconductor in the ground state. Suppose that, as a result of an external effect, one of the electrons from the pair $(q, -q)$ is moved to a neighboring state in momentum space. Recall that initially all states were either occupied in pairs or empty in pairs. Then the transfer of one electron from the pair $(q, -q)$ to the neighboring state implies that two uncoupled (excited) electrons have appeared. One of them stays in one of the states $(q, -q)$ while the other turns up in the neighboring $\boldsymbol{k}$ state and, naturally, does not have a partner in the $-\boldsymbol{k}$ state. But we have just found out that, in order to break a pair, one needs at least the energy $2\Delta_0$, as illustrated by the diagram in Fig. 6.6. All pair states belong to the condensate occupying the ground-state energy level. A single extra electron is not allowed to be at this level and must therefore occupy the first empty level available in the elementary excitation spectrum. If a pair is broken, both electrons must go up to the elementary excitation levels which requires an energy larger than $2\Delta_0$.

The existence of this energy gap is a very important property of the superconductor and explains many aspects of its behavior.

6.3.2 Density of States of Elementary Excitations in the Superconductor and the Coherence Length

The lower part of the elementary excitation spectrum is illustrated in Fig. 6.6. An expression for the density of states in this spectrum can be easily derived from (6.11). Indeed, $E_{\boldsymbol{k}}$ is the energy of an elementary excitation (see (6.24)), that is, it tells us how much the energy of the system increases when an extra electron with momentum $\boldsymbol{k}$ is added to the superconductor:

$$E_k = \sqrt{\varepsilon_k^2 + \Delta_0^2} = \sqrt{\left(\frac{\hbar^2 k^2}{2m} - \frac{\hbar^2 k_F^2}{2m}\right)^2 + \Delta_0^2} .$$

The dependence of E_k on k given by this expression is shown in Fig. 6.7.

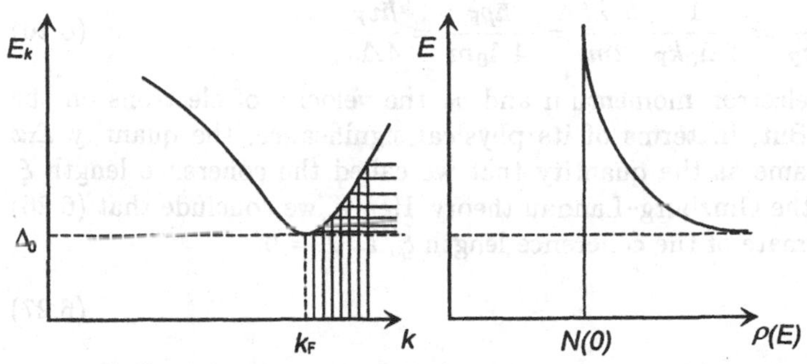

Fig. 6.7. Spectrum of elementary excitations, E_k, of the superconductor and the density of states $\rho(E)$

It is easy to see from the plot that the energy levels of elementary excitations become denser at $E_k \to \Delta_0$. The density of states or, in other words, the number of the energy levels per unit energy interval and unit volume (1 cm^3) of the material, is

$$\rho(E) = \mathrm{d}\nu/\mathrm{d}E ,$$

where $\mathrm{d}\nu$ is the number of energy levels in the energy interval $\mathrm{d}E$, in the vicinity of the level E. We recall that $\mathrm{d}\nu/\mathrm{d}\varepsilon$ gives the density of states near the Fermi level for the normal metal, i.e.,

$$\mathrm{d}\nu/\mathrm{d}\varepsilon = N(0) .$$

Therefore

$$\rho(E) = \frac{\mathrm{d}\nu}{\mathrm{d}\varepsilon}\frac{\mathrm{d}\varepsilon}{\mathrm{d}E} = N(0)\frac{E}{\sqrt{E^2 - \Delta_0^2}} . \qquad (6.25)$$

It follows from (6.25) that, at $E \to \Delta_0$, the density of states of elementary excitations in the superconductor is indeed $\rho(E) \to \infty$, as illustrated in Fig. 6.7.

Let us now discuss how the superconducting coherence length can be evaluated using the microscopic theory. The wavefunction of the ground state of the superconductor is already known to us from Sect. 6.2.1. The ground state can be represented by the distribution of electron pairs in momentum space given by the function v_k^2. The dependence of v_k^2 on k is plotted in Fig. 6.5. One can see that large variations of v_k^2 can occur only within the region

$$\Delta k \sim 2\Delta_0 \frac{k_F}{\varepsilon_F}.$$

Then, in real space, large variations of the ground state wavefunction can be expected within the interval Δx defined by the relation

$$\Delta x \Delta k \sim 1.$$

It follows that

$$\Delta x \sim \frac{\varepsilon_F}{2\Delta_0 k_F} = \frac{1}{2\Delta_0 k_F} \frac{\hbar^2 k_F^2}{2m} = \frac{\hbar p_F}{4\Delta_0 m} = \frac{\hbar v_F}{4\Delta_0}. \tag{6.26}$$

Here p_F is the electron momentum and v_F the velocity of electrons on the Fermi surface. But, in terms of its physical significance, the quantity Δx is exactly the same as the quantity that we called the coherence length ξ, when studying the Ginzburg–Landau theory. Hence, we conclude that (6.26) provides an estimate of the coherence length ξ_0 at $T = 0$:

$$\xi_0 \sim \frac{\hbar v_F}{4\Delta_0}. \tag{6.27}$$

It will be demonstrated below that $\Delta_0 \sim k_B T_c$, where k_B is the Boltzmann constant and T_c is the critical temperature. A rigorous calculation yields

$$\xi_0 = 0.18 \frac{\hbar v_F}{k_B T_c}. \tag{6.28}$$

The quantity ξ_0 can also be treated as the size of a Cooper pair which can be evaluated as

$$\xi_0 \simeq 0.18 \frac{10^{-27} \times 10^8}{1.38 \times 10^{-16} \times 1} \sim 10^{-4}\ \text{cm},$$

where $v_F \sim 10^8\ \text{cm}\,\text{s}^{-1}$ and $T_c \sim 1$ K. This is a large, macroscopic distance.

6.3.3 Temperature Dependence of the Energy Gap

As the temperature increases, the energy gap Δ decreases (the notation Δ_0 will be kept for $T = 0$). This is easy to understand. As we already know, in order to break a Cooper pair and create two elementary excitations, the energy 2Δ is needed. If the temperature T is such that $k_B T \sim 2\Delta$, it is evident that many Cooper pairs will be broken through thermal processes. Accordingly, a large number of states in momentum space will be filled by elementary excitations (single electrons). But this implies that these states can no longer participate in the pair transitions and, therefore, cannot contribute to the net reduction of the superconductor's energy. Consequently, the energy of the superconductor has to increase. These states will likewise be unable to participate in forming the energy gap (see (6.10)). Hence, as the number of broken pairs grows, there is an increase of the number of elementary excitations and a decrease of the energy gap.

Let us analyze these processes quantitatively.

Since the elementary excitations obey the Fermi–Dirac statistics, the probability that the state $\boldsymbol{k}$ is occupied by a single electron is

$$f_{\boldsymbol{k}} = \frac{1}{\exp\left(\frac{E_{\boldsymbol{k}}}{k_{\mathrm{B}}T}\right) + 1} , \tag{6.29}$$

where $E_{\boldsymbol{k}}$ is the energy of an elementary excitation. One can see that $f_{\boldsymbol{k}} \ll 1$ at $k_{\mathrm{B}}T \ll E_{\boldsymbol{k}}$, and $f_{\boldsymbol{k}} \simeq 0.5$ at $k_{\mathrm{B}}T \gg E_{\boldsymbol{k}}$. If at least one of the states, $(\boldsymbol{k})$ or $(-\boldsymbol{k})$, is occupied, the pair state $(\boldsymbol{k}, -\boldsymbol{k})$ cannot take part in creating the superconducting state. The probability of this is $2f_{\boldsymbol{k}}$. Hence, the probability that the pair state $(\boldsymbol{k}, -\boldsymbol{k})$ can participate in the scattering processes, i.e., can take part in creating the superconducting state, is $1 - 2f_{\boldsymbol{k}}$.

The expression for the total energy of the superconductor at $T \neq 0$ can then be written in the form (see Sect. 6.2)

$$\begin{aligned} W &= \sum_{\boldsymbol{k}} 2\,|\varepsilon_{\boldsymbol{k}}|\, f_{\boldsymbol{k}} + 2\sum_{\boldsymbol{k}} \varepsilon_{\boldsymbol{k}}(1 - 2f_{\boldsymbol{k}})\, v_{\boldsymbol{k}}^2 \\ &\quad - V{\sum_{\boldsymbol{k}\,\boldsymbol{k}'}}' v_{\boldsymbol{k}} u_{\boldsymbol{k}} v'_{\boldsymbol{k}} u'_{\boldsymbol{k}}\,(1 - 2f_{\boldsymbol{k}})\,(1 - 2f_{\boldsymbol{k}'}) \,. \end{aligned} \tag{6.30}$$

Here the first term is the kinetic energy of the elementary excitations, the second term is the kinetic energy of the superconducting electrons and the last term is the energy of the phonon-mediated electron interaction which is the prime cause of the superconducting state. The two last factors account for the probability of this interaction.

The free energy density of the superconductor is

$$F = W - TS \,, \tag{6.31}$$

where S is the entropy of the material. The functions $v_{\boldsymbol{k}}^2$, describing the distribution of the supeconducting electrons in momentum space at thermodynamic equilibrium, can be found from the condition that the free energy density F is a minimum:

$$\frac{\partial F}{\partial (v_q^2)} = 0 \,. \tag{6.32}$$

Substituting (6.31) and (6.30) into (6.32) yields

$$\frac{v_q u_q}{1 - 2v_q^2} = \frac{\Delta}{2\varepsilon_q} \,, \tag{6.33}$$

where

$$\Delta = V{\sum_{\boldsymbol{k}}}' v_{\boldsymbol{k}} u_{\boldsymbol{k}}\,(1 - 2f_{\boldsymbol{k}}) \,. \tag{6.34}$$

This last expression gives the temperature dependence of the energy gap. As $T \to 0$, the gap is $\Delta \to \Delta(0)$, where $\Delta(0) = \Delta_0$ is the gap at $T = 0$ defined in the preceding section (see (6.10)).

From (6.34) we can derive the equation for the energy gap. First, by analogy with (6.12), we write v_q^2 in the form

$$v_q^2 = \frac{1}{2}\left(1 - \frac{\varepsilon_q}{E_q}\right) ,$$

where

$$E_q = \sqrt{\varepsilon_q^2 + \Delta^2(T)} .$$

Then (6.34) reduces to

$$\Delta = V {\sum_k}' \frac{\Delta}{2E_k}\left(1 - \frac{2}{\exp\left(E_k/k_\mathrm{B}T\right) + 1}\right) .$$

Replacing summation with integration yields, after simple algebra,

$$\frac{1}{N(0)\,V} = \int_0^{\hbar\omega_\mathrm{D}} \frac{\mathrm{d}\varepsilon}{\sqrt{\varepsilon^2 + \Delta^2(T)}} \tanh \frac{\sqrt{\varepsilon^2 + \Delta^2(T)}}{2k_\mathrm{B}T} . \tag{6.35}$$

Thus we have obtained an implicit temperature dependence of the energy gap. This dependence is sketched in Fig. 6.8: Near T_c, the variation of the gap with temperature obeys $\Delta \propto (T_\mathrm{c} - T)^{1/2}$.

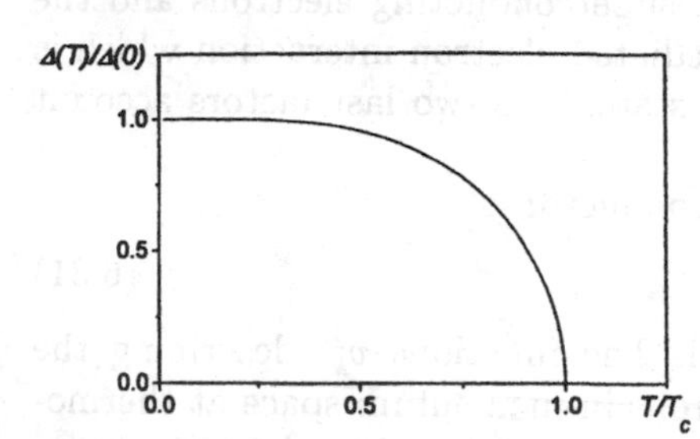

Fig. 6.8. Temperature dependence of the energy gap in the BCS theory

From (6.35), an explicit expression for the critical temperature T_c can also be derived. Indeed, at $T = T_\mathrm{c}$, the gap is $\Delta = 0$. Hence, replacing T in (6.35) with T_c and setting $\Delta = 0$ yields an equation with respect to T_c:

$$\frac{1}{N(0)\,V} = \int_0^{\hbar\omega_\mathrm{D}} \frac{\mathrm{d}\varepsilon}{\varepsilon} \tanh \frac{\varepsilon}{2k_\mathrm{B}T_\mathrm{c}} . \tag{6.36}$$

Carrying out the integration, we get

$$k_\mathrm{B}T_\mathrm{c} = 1.14\,\hbar\omega_\mathrm{D} \exp\left(-\frac{1}{N(0)\,V}\right) . \tag{6.37}$$

On the other hand, we already know (see (6.16)) that

$$\Delta_0 = 2\,\hbar\omega_\mathrm{D} \exp\left(-\frac{1}{N(0)\,V}\right) .$$

Then

$$2\Delta_0 = 3.52\, k_B T_c\ . \tag{6.38}$$

These relations are in good quantitative agreement with numerous experiments.

Let us outline once again the physical significance of the results just obtained. Due to thermal processes, some of the Cooper pairs become broken and turn into uncoupled single electrons which can also be classified as normal electrons or elementary excitations. These are simply different names for the same physical object. The single electrons occupy certain $\boldsymbol{k}$ states in momentum space thereby excluding these states (and their partner states) from the sum in (6.10) which defines the gap Δ. As a result, the gap shrinks. When it finally becomes zero, the superconductor goes to the normal state. This condition determines the critical temperature – see (6.37).

Remarkably, (6.37) also provides an explanation for the isotope effect: For different isotopes of the same superconducting element, the critical temperatures are different and follow the rule

$$T_c M^{1/2} = \text{const}\ ,$$

where M is the mass of the isotope. Recalling that the Debye frequency varies as

$$\omega_D \propto M^{-1/2}\ ,$$

we immediately obtain the isotope effect from (6.37).

6.4 Tunneling Effects in Superconductors

The most direct way of examining the energy gap is by tunneling experiments. This technique was developed by Giaever in 1960 [63]. Its principle is very simple (see Fig. 6.9). A narrow film of the first metal is deposited onto a glass substrate with contact pads made in advance. Then the film is oxidized, so that it is covered with an insulating oxide layer of several nanometers thickness (the barrier layer). After that, a narrow film of the second metal is deposited in the transverse direction. The intersection of the two films (of area $\sim 1\,\text{mm}^2$) forms a tunnel junction. The experiment consists in examining the current–voltage characteristic of this junction. Physically, the experiment employs the special quantum-mechanical ability of electrons to pass through thin potential barriers (in our case, through the insulating layer) by means of tunneling.

Let us discuss the current–voltage characteristics expected for different types of tunnel junctions.

The simplest case is when both the first and the second metal are normal. As soon as the two metals are in contact, their Fermi levels become equal. The corresponding energy-level diagram is shown in Fig. 6.10 (a). If a voltage

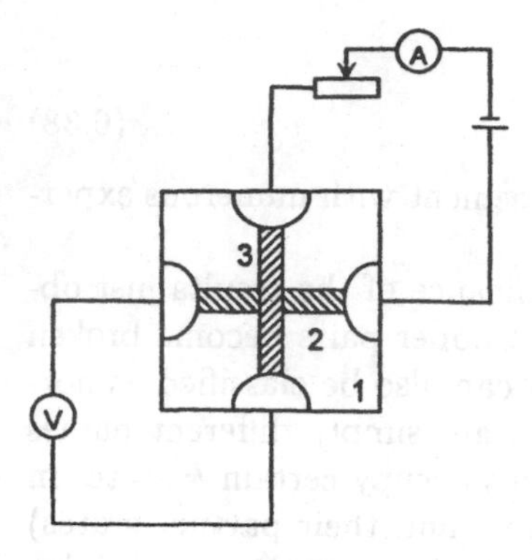

Fig. 6.9. Scheme of an electron tunneling experiment: *1* – glass substrate; *2* – thin film of the first metal; *3* – thin film of the second metal

V is then applied to the junction (see Fig. 6.10 (b)), the Fermi levels will shift apart by the amount of eV and a tunneling current will begin to flow. If the density of states in the energy interval of interest is assumed to be constant, the current will be directly proportional to eV, as is evident from Fig. 6.10 (b). In other words, the current will obey Ohm's law.

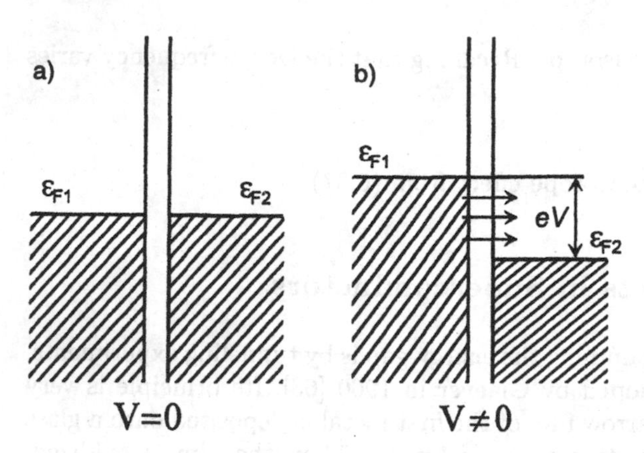

Fig. 6.10. Energy-level diagram for two normal metals brought into contact: **(a)** $V = 0$; **(b)** $V \neq 0$, a tunneling current proportional to eV passes through the junction

Let us now consider the case when one of the metals is normal and the other superconducting. The elementary excitation spectrum of the superconductor is already known to us (see Figs. 6.6 and 6.7). The energy of elementary excitations is measured from the ground-state level of the superconductor. In the normal metal, the energy of elementary excitations is measured from the Fermi level. Hence, when a tunnel junction is established between a normal metal and a superconductor, the energy levels to be equalized (due to the voltage drop) are the Fermi level of the normal metal and the ground-state level of the superconductor, as illustrated in Fig. 6.11 (a).

Let us apply a voltage V to the junction. Suppose that, as a result, the Fermi level of the normal metal has moved up by the value of eV above the ground-state level of the superconductor. A current from N to S can flow only when $eV \geq \Delta$, as is evident from Fig. 6.11 (b), where Δ is the energy

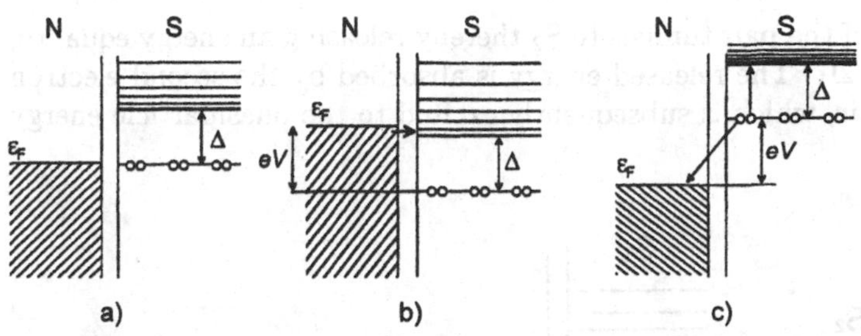

Fig. 6.11. Energy-level diagram for an NS tunnel junction: **(a)** $V = 0$, ε_F coincides with the ground-state level of the superconductor; **(b)** $V \neq 0$, $|eV| > \Delta$, a finite tunneling current passes from N to S; **(c)** $V \neq 0$, $|eV| > \Delta$, a finite tunneling current passes from S to N

gap of the superconductor. The result will be the same if the polarity of the applied voltage is reversed. In that case, electrons will tunnel from S to N and a current will start to flow when the break-up of Cooper pairs in S becomes possible. That is, the energy released as a result of the tunneling of one electron from the pair has to be sufficient for the second electron to be excited into the energy band allowed for single electrons, i.e., above the energy gap (see Fig. 6.11 (c)).

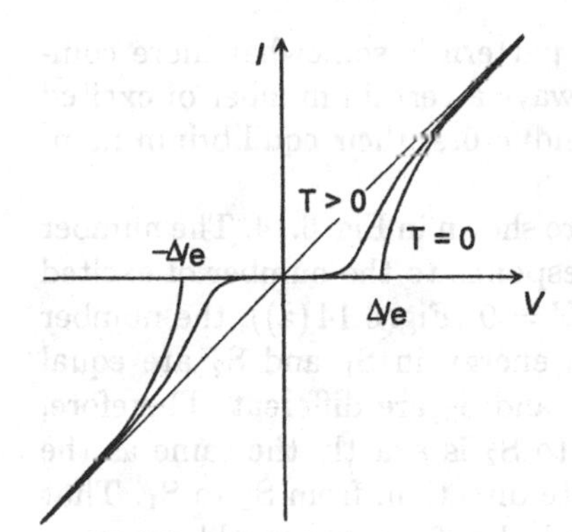

Fig. 6.12. Current–voltage characteristic of an NS tunnel junction

Thus the current–voltage characteristic of such a junction must be an odd function of V, as illustrated in Fig. 6.12. In the case $T \neq 0$, the current–voltage characteristic will be somewhat smeared out, as one would expect.

Let us turn to the tunneling effects which occur when both metals (S_1 and S_2) are superconducting. Then the cases $T = 0$ and $T \neq 0$ are essentially different.

Consider the case $T = 0$ first. It follows from Fig. 6.13 that a tunneling current can start to flow only when the voltage across the junction is $V > (\Delta_1 + \Delta_2)/e$. Indeed, it is only on this condition that the tunneling process shown in Fig. 6.13 becomes possible: a Cooper pair in S_1 breaks up and one of the electrons of the pair tunnels to S_2 thereby releasing an energy equal to, or greater than, Δ_1. The released energy is absorbed by the second electron of the broken pair, which is subsequently excited to the quasiparticle energy band of S_1.

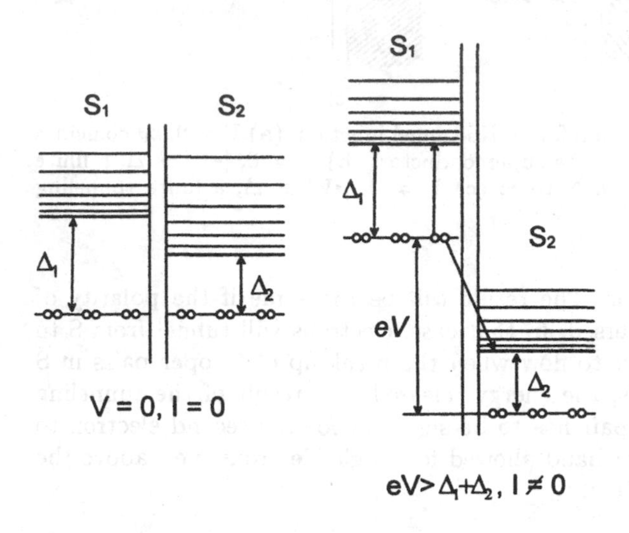

Fig. 6.13. Energy-level diagrams for an S_1S_2 tunnel junction at $T = 0$

In a real experiment, when $T \neq 0$, the pattern is somewhat more complicated. At finite temperatures, there is always a certain number of excited single (unpaired) electrons in both superconductors; their equilibrium number is determined by the temperature.

The energy-level diagrams for this case are shown in Fig. 6.14. The number of dots at a given energy level roughly corresponds to the number of excited states at this energy. One can see that, at $V = 0$ (Fig. 6.14 (a)), the number of excited states at the levels of the same energy in S_1 and S_2 are equal despite the fact that the energy gaps in S_1 and S_2 are different. Therefore, the number of particles tunneling from S_1 to S_2 is exactly the same as the number of particles tunneling in the opposite direction, from S_2 to S_1. That is, at equilibrium, the net tunneling current is $I = 0$, as one would expect.

If we now apply a voltage V, the equilibrium will be immediately lifted and a flow of quasiparticles will start from one superconductor to the other, even if the voltage is very small. One should bear in mind, however, that the quasiparticle density of states in the superconductor has a singularity at $E = \Delta$ (see (6.25) and Fig. 6.7). Hence, if the applied voltage is precisely $eV = \Delta_1 - \Delta_2$ (see Fig. 6.14 (b)), the equalized energy levels in S_1 and S_2 will be those having a density of states $\rho = \infty$. This will result in a sharp rise of the tunneling current. As the voltage V is increased further, the energy levels with $\rho = \infty$ will move apart again and the tunneling current will decrease. This implies that the current–voltage characteristic exhibits a maximum at $V = (\Delta_1 - \Delta_2)/e$, as illustrated in Fig. 6.15. If the applied voltage is reversed,

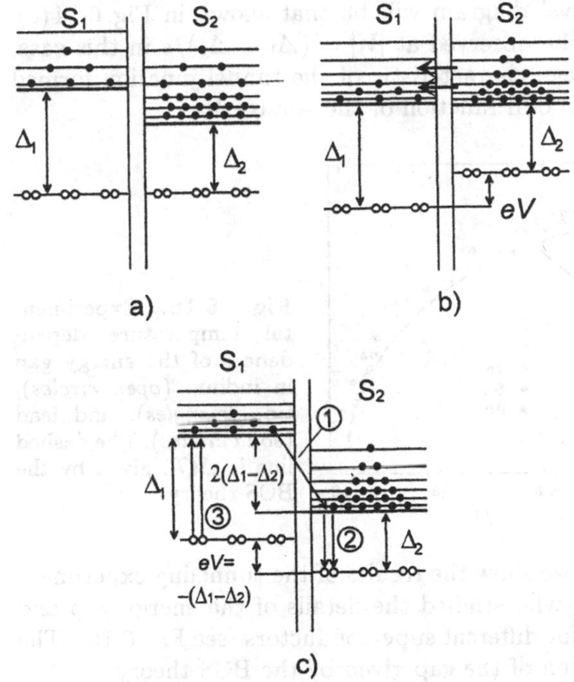

Fig. 6.14. Tunneling between two superconductors at $T \neq 0$: **(a)** $V = 0$, the numbers of excitations of the same energy are identical in S_1 and S_2 and the current is zero; **(b)** $eV = \Delta_1 - \Delta_2$, the finite current is due to the transfer of excited particles from S_1 to S_2; **(c)** $eV = -(\Delta_1 - \Delta_2)$, an excited particle passes from S_2 to S_1 (process 1); in S_2, it couples with one of the electrons there and relaxes back to the ground level (process 2); the energy $2\Delta_1$ released during process 2 is sufficient to break a Cooper pair in S_1 (process 3)

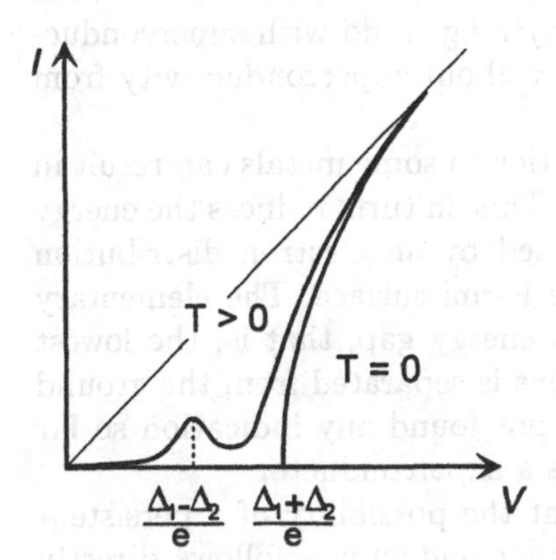

Fig. 6.15. Current–voltage characteristic of an S_1S_2 tunnel junction

the corresponding energy-level diagram will be that shown in Fig. 6.14 (c). A sharp rise of current will be observed at $|V| = (\Delta_1 - \Delta_2)/e$ in this case, too. Hence the current–voltage characteristic of the tunnel junction formed by two superconductors is an odd function of the voltage V.

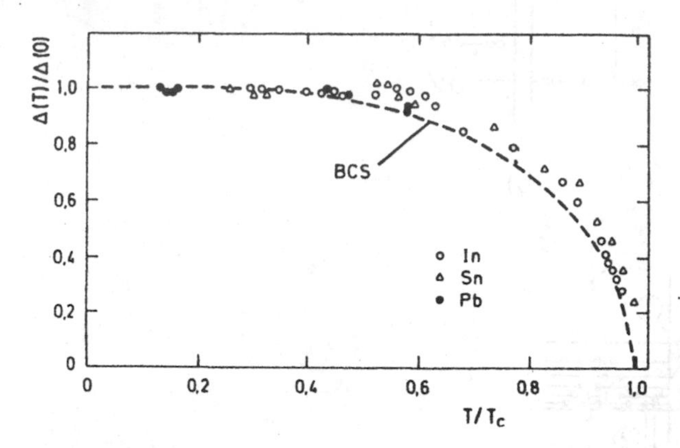

Fig. 6.16. Experimental temperature dependences of the energy gap in indium (*open circles*), tin (*triangles*), and lead (*solid circles*). The dashed line is $\Delta(T)$ given by the BCS theory

To conclude this section, we show the results of the tunneling experiment by Giaever and Megerle [64] who studied the details of the energy gap and its temperature dependence for different superconductors (see Fig. 6.16). The dashed line shows the variation of the gap given by the BCS theory.

6.5 Persistent Current and the Meissner–Ochsenfeld Effect

Ealier in this chapter, we discussed the basic ideas of the superconducting state. We gave a description of the ground state of the superconductor, examined the elementary excitation spectrum, and proved the existence of the energy gap. However, the reader may have wanted to ask whether there is any proof that everything said so far has anything to do with superconductivity. After all, at this moment, all we know about superconductivity from the microscopic theory is only the following.

The existence of electron–phonon interaction in some metals can result in an effective attraction between the electrons. This, in turn, reduces the energy of the ground state which is then represented by an electron distribution smeared out in momentum space about the Fermi surface. The elementary excitation spectrum of such a metal has an energy gap, that is, the lowest energy level allowed for elementary excitations is separated from the ground state by the energy gap. However, we have not found any indication so far that a material having all these properties is a superconductor.

In this section, we shall demonstrate that the possibility of a persistent current – a fundamental property of the superconductor – follows directly from the existence of the energy gap. For a better understanding, let us first examine the normal metal. In particular, we are interested in the electron distribution in momentum space in the presence of a constant current passing through a normal metal. At $T = 0$ and zero current, all electrons of the metal occupy the states within the Fermi surface while all states outside it are empty. If we then apply an electric field to the metal (for example, in the direction of the x axis), the electrons will start accelerating in this direction. But accelerated motion of electrons in real space is equivalent to their steady motion in momentum space, with constant velocity. This means that the Fermi sphere as a whole will start to move at a steady rate in the direction of the k_x axis in momentum space. This steady motion can be considered uninterrupted as long as electron collisions with crystal defects or impurity atoms can be neglected. When this is no longer possible, there arises a dynamic equilibrium: As a resultof scattering, the electrons having the maximum value of momentum component k_x are carried to the empty states in momentum space until an equilibrium is attained. This implies that, despite the electric field and the steady motion of electrons in momentum space in the direction of k_x, their scattering results in an electron distribution in momentum space which is on average stationary. The Fermi surface shifts somewhat with respect to the origin and the electron scattering results in a transfer of energy to the crystal lattice, that is, in heating up the conductor.

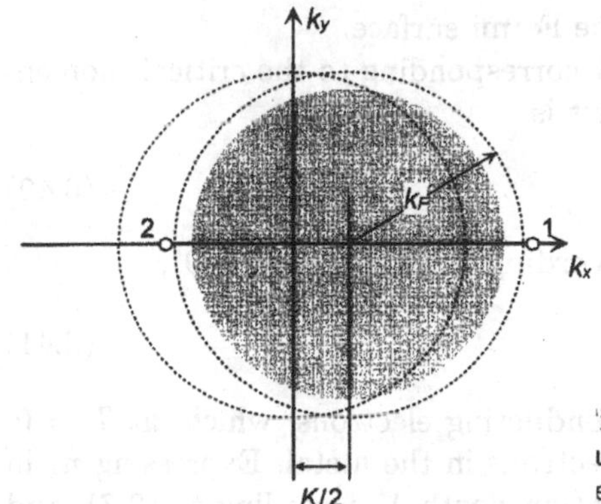

'ig. 6.17. Electron states in a superconuctor carrying a current. The 'smeared' ermi sphere shifts as a whole by $K/2$

Consider now the superconductor. In this case, a current can start to flow (in the absence of an electric field!) if all Cooper pairs have the same momentum $\boldsymbol{K}$. Suppose that the direction of the current coincides with the x axis, that is, $\boldsymbol{K} = (K, 0, 0)$. This is equivalent to a shift of the 'smeared' Fermi sphere in momentum space by the value of $K/2$ in the direction of the k_x axis, as illustrated in Fig. 6.17. Let us follow the transformations of a Cooper pair (*1,2*) of momentum $(k_\mathrm{F} + K/2,\ 0,\ 0)$ indicated in the figure. Electron *1* having the maximum kinetic energy $(\hbar^2/2m)\,(k_\mathrm{F} + K/2)^2$ would have benefited from a transition to an empty state somewhere near electron

2. Such a process would have been accompanied by an energy decrease by the value of

$$\frac{\hbar^2}{2m}\left(k_\mathrm{F}+\frac{K}{2}\right)^2-\frac{\hbar^2}{2m}\left(k_\mathrm{F}-\frac{K}{2}\right)^2=\frac{\hbar^2}{m}k_\mathrm{F}K\,,$$

where m is the electron mass and k_F is the radius of the Fermi sphere. But then the pair (*1*, *2*) must be broken, which would cause an increase of the total energy of the system by 2Δ, where Δ is the energy gap.

It is now evident that, at sufficiently small currents (or sufficiently small K), the energy gain $\hbar^2 k_\mathrm{F}K/m$ cannot compensate for the energy loss 2Δ and the pair will not break up. Of course, on the whole, such a state is less favored energetically than the state without a current ($K=0$), but it can happen to be metastable and exist for an infinitely long time. An example of such a state is provided by a closed superconducting ring carrying a current: A break-down of the current state begins when it becomes energetically favorable to break Cooper pairs, that is, when

$$\frac{\hbar^2 k_\mathrm{F}K_\mathrm{c}}{m}\approx 2\Delta\,,$$

where $\hbar K_\mathrm{c}=P_\mathrm{c}$ is the critical momentum of a pair. Hence

$$P_\mathrm{c}=\hbar K_\mathrm{c}\approx\frac{2m\Delta}{p_\mathrm{F}}=\frac{2\Delta}{v_\mathrm{F}}\,,\tag{6.39}$$

where v_F is the electron velocity on the Fermi surface.

Let us evaluate the current density corresponding to the critical momentum. Since the critical velocity of a pair is

$$v_\mathrm{c}=\frac{P_\mathrm{c}}{2m}\,,\tag{6.40}$$

the critical current density will be, according to (6.39) and (6.40),

$$j_\mathrm{c}=n_\mathrm{s}ev_\mathrm{c}=n_\mathrm{s}e\,\frac{\Delta}{mv_\mathrm{F}}\,,\tag{6.41}$$

where n_s is the density of the superconducting electrons, which, at $T=0$, is equivalent to the total density of electrons in the metal. Expressing n_s in terms of the London magnetic penetration depth λ, according to (2.7), and Δ_0 in terms of the coherence length ξ, according to (6.27), we obtain the critical current density at $T=0$ in the form

$$j_\mathrm{c}=\frac{\hbar c^2}{16\pi e}\,\frac{1}{\lambda^2\xi_0}\,.\tag{6.42}$$

To estimate the order of magnitude of j_c, we assume $\lambda\sim 10^{-5}$ cm and $\xi_0\sim 10^{-4}$ cm. Then

$$j_\mathrm{c}\sim\frac{10^{-27}\,(3\times10^{10})^2}{16\pi\times4.8\times10^{-10}}\,\frac{1}{(10^{-5})^2\,10^{-4}}\sim 4\times10^{15}\ \text{CGS units}\,.$$

Recalling that 1 A is equivalent to 3×10^9 CGS units of current, we have

$$j_c = \frac{4 \times 10^{15}}{3 \times 10^9} \approx 1.3 \times 10^6 \text{ A cm}^{-2} .$$

The critical current density j_c in (6.41) can be expressed in terms of $H_{cm}(0)$ and λ. Using the expressions for the density of states near the Fermi surface, $N(0) = k_F m/(\pi^2 \hbar^2)$, and the electron density $n_s = k_F^3/3\pi^2$, together with (6.20) and (6.27), we obtain

$$j_c = \frac{1}{\sqrt{3}\, 4\pi} \frac{c H_{cm}(0)}{\lambda} . \tag{6.43}$$

It is interesting to compare the result (6.43) with the critical current density (3.67) given by the Ginzburg–Landau theory. One can easily verify that the two results are equivalent except for a numerical factor (if, of course, (6.43) is extrapolated to temperatures close to T_c). The discrepancy in the numerical factor is not surprising because (6.43) is only intended as an estimate.

Thus we can draw a very important conclusion: the critical current (3.67) given by the Ginzburg–Landau theory has the physical significance of a pair-breaking current, that is, of a current that causes electron pairs in the condensate to become unstable.

Let us now turn to a discussion of how the existence of the energy gap in the elementary excitation spectrum of the superconductor explains the Meissner–Ochsenfeld effect, i.e., the exclusion of weak magnetic fields from the interior of the superconductor. Rigorous calculations which prove that the existence of the energy gap is a sufficient condition for the Meissner effect are beyond the scope of this book. Nevertheless, we shall attempt an interpretation of the Meissner effect using the language of microscopic theory.

Following London [65], one can introduce the so-called 'rigidity' of the electron wavefunction of the superconductor. This means that, owing to the energy gap and the coherent behavior of the whole body of electrons, a weak magnetic field is unable to change the wavefunction in any significant way. From this the Meissner effect follows immediately.

Indeed, assume that $\Psi(\boldsymbol{r}_1, \boldsymbol{r}_2, \ldots, \boldsymbol{r}_N)$ is the wavefunction of N electrons in the superconductor. Note that here Ψ is not an effective wavefunction of the superconducting electrons, as in the Ginzburg–Landau theory (which depends on only one coordinate $\boldsymbol{r}$), but a true wavefunction. Then the following expression for a current in the presence of a magnetic field can be derived using general relations of quantum mechanics:

$$\begin{aligned} \boldsymbol{j}(\boldsymbol{r}) \;=\; & \sum_{n=1}^{N} \int \cdots \int \left\{ \frac{e\hbar}{2im} \left[\Psi^* \nabla_n \Psi - \Psi \nabla_n \Psi^*\right] \right. \\ & \left. - \frac{e^2}{mc} \boldsymbol{A}(\boldsymbol{r}_n) \, |\Psi|^2 \right\} \delta(\boldsymbol{r} - \boldsymbol{r}_n) \, \mathrm{d}^3\boldsymbol{r}_1 \ldots \mathrm{d}^3\boldsymbol{r}_N . \end{aligned} \tag{6.44}$$

Here $\boldsymbol{A}$ is the vector potential of the magnetic field. If the magnetic field is zero ($\boldsymbol{A} = 0$), the current and the integral of the first term are zero, too:

$$\sum_{n=1}^{N} \int \dots \int [\Psi^* \nabla_n \Psi - \Psi \nabla_n \Psi^*] \, \delta(\boldsymbol{r} - \boldsymbol{r}_n) \mathrm{d}^3 \boldsymbol{r}_1 \dots \mathrm{d}^3 \boldsymbol{r}_N = 0 \,. \tag{6.45}$$

But we have assumed that the wavefunction is 'rigid', that is, it is not affected by the magnetic field. Hence, formula (6.45) must hold in a nonzero magnetic field. Then, omitting the first term of (6.44), we obtain, after elementary integration

$$\boldsymbol{j}(\boldsymbol{r}) = -\mathrm{const} \times \boldsymbol{A}(\boldsymbol{r}) \,. \tag{6.46}$$

This last formula is equivalent to the well-known London equation (2.17) and, therefore, is indeed a mathematical expression of the Meissner effect.

6.6 Relation Between the Microscopic and the Ginzburg–Landau Theory

The Ginzburg–Landau (GL) theory is, as we already know, phenomenological, that is, it does not allow its coefficients to be interpreted on the microscopic level. A detailed interpretation of the quantities involved in the GL theory became possible only after the rigorous microscopic theory of superconductivity had been developed by Bardeen, Cooper, and Schrieffer. This task was accomplished by Gorkov [66]. Here we shall only outline the final results, which differ for 'clean' and 'dirty' superconductors. In what follows, all quantities corresponding to 'clean' superconductors will be labeled with an index 'p' and those corresponding to 'dirty' superconductors with an index 'd'. A superconductor is considered clean if $l \gg \xi_0$, where l is the electron mean free path. A dirty superconductor is characterized by $l \ll \xi_0$.

As one should expect, the energy gap $\Delta(T)$ represents the order parameter in the superconductor and, therefore, at $T \to T_\mathrm{c}$, must be proportional to the order parameter Ψ of the GL theory. Indeed, the exact result is:

$$\Psi_\mathrm{p} = \left[\frac{7\zeta(3)\, m v_\mathrm{F}^2\, N(0)}{2\pi^2 k_\mathrm{B}^2 T_\mathrm{c}^2} \right]^{1/2} \Delta(\boldsymbol{r}) \,, \quad \zeta(3) \approx 1.202 \,,$$

$$\Psi_\mathrm{d} = \left[\frac{\pi m v_\mathrm{F}\, N(0)\, l}{12 \hbar k_\mathrm{B} T_\mathrm{c}} \right]^{1/2} \Delta(\boldsymbol{r}) \,,$$

where $\zeta(x)$ is the Riman function.

At $T \to T_\mathrm{c}$ we have

$$\Delta(T) = \left(\frac{8\pi^2}{7\zeta(3)} \right)^{1/2} k_\mathrm{B} T_\mathrm{c} \left(1 - \frac{T}{T_\mathrm{c}} \right)^{1/2} ,$$

or

$$\Delta(T) \approx 3.1\, k_\mathrm{B} T_\mathrm{c} \left(1 - \frac{T}{T_\mathrm{c}} \right)^{1/2} .$$

The coherence length and the magnetic penetration depth are, accordingly,

$$\xi_{\rm p} = 0.74\,\xi_0\left(1-\frac{T}{T_{\rm c}}\right)^{-1/2},$$

$$\xi_{\rm d} = 0.85\,(\xi_0 l)^{1/2}\left(1-\frac{T}{T_{\rm c}}\right)^{-1/2},$$

$$\lambda_{\rm p} = \frac{\lambda(0)}{\sqrt{2}}\left(1-\frac{T}{T_{\rm c}}\right)^{-1/2},$$

$$\lambda_{\rm d} = 0.615\,\lambda(0)\left(\frac{\xi_0}{l}\right)^{1/2}\left(1-\frac{T}{T_{\rm c}}\right)^{-1/2},$$

$$\lambda^2(0) = \frac{3c^2}{8\pi e^2 v_{\rm F}^2 N(0)}\,.$$

The GL coefficients α and β in the expansion of the free energy in powers of the order parameter Ψ are:

$$\alpha_{\rm p} = 1.83\,\frac{\hbar^2}{2m}\frac{1}{\xi_0{}^2}\left(\frac{T}{T_{\rm c}}-1\right),$$

$$\alpha_{\rm d} = 1.36\,\frac{\hbar^2}{2m}\frac{1}{\xi_0 l}\left(\frac{T}{T_{\rm c}}-1\right),$$

$$\beta_{\rm p} = 0.35\,\frac{1}{N(0)}\left(\frac{\hbar^2}{2m\xi_0{}^2}\right)^2\frac{1}{(k_{\rm B}T_{\rm c})^2},$$

$$\beta_{\rm d} = 0.2\,\frac{1}{N(0)}\left(\frac{\hbar^2}{2m\xi_0 l}\right)^2\frac{1}{(k_{\rm B}T_{\rm c})^2}\,.$$

The GL parameter takes the values

$$\kappa_{\rm p} = 0.96\,\frac{\lambda(0)}{\xi_0}\,,\qquad \kappa_{\rm d} = 0.725\,\frac{\lambda(0)}{l}\,.$$

Let us now establish the range of validity for the GL theory. In the series expansion (3.7) of the Gibbs free energy density $G_{{\rm s}H}$ in powers of $|{\rm i}\hbar\nabla\Psi-(2e/c)\boldsymbol{A}\Psi|^2$, only the first term has been kept. This means that only slow changes of Ψ and $\boldsymbol{A}$ are assumed over distances comparable with the characteristic size of an inhomogeneity in the superconductor, that is, over the size of the Cooper pair.

In the case of a clean superconductor, when the electron mean free path satisfies the inequality $l \gg \xi_0$, the GL theory is valid if $\xi(T)$, $\lambda(T) \gg \xi_0$. Since $\xi(T) \sim \xi_0\,(1-T/T_{\rm c})^{-1/2}$, the quantity $\xi(T)$ always exceeds ξ_0 at $T \sim T_{\rm c}$ and the first condition of validity is satisfied automatically.

The second condition, $\lambda(T) \gg \xi_0$, represents the requirement that local electrodynamics is applicable, or, in other words, that the superconductor is of the London type. Since $\lambda(T) \sim \lambda(0)\,(1-T/T_{\rm c})^{-1/2}$ and $\kappa \sim \lambda(0)/\xi_0$, the condition $\lambda(T) \gg \xi_0$ reduces to

$$\kappa^2 \gg 1 - \frac{T}{T_c}\,. \tag{6.47}$$

This is a rather strict condition because κ in type-I superconductors can assume a small value. For example, $\kappa\,(\mathrm{Al}) = 0.01$, $\kappa\,(\mathrm{Pb}) = 0.23$. In 'dirty' superconductors ($l \ll \xi_0$), the validity interval for the GL theory is much wider. In this case, the characteristic scale of inhomogeneity is the mean free path l and the GL theory can be applied if $\xi(T)$, $\lambda(T) \gg l$. Since $\xi(T) \sim (\xi_0 l)^{1/2}\,(1 - T/T_c)^{-1/2}$, the condition $\xi(T) \gg l$ reduces to $\xi_0/l \gg 1 - T/T_c$. Furthermore, since $\xi_0 \gg l$, this condition is much less strict than the general condition of validity for the Landau theory of the second-order phase transitions, $T_c - T \ll T_c$.

Consider the second condition: $\lambda(T) \gg l$. Recalling that, for dirty superconductors, $\lambda(T) \sim \lambda(0)\,(\xi_0/l)^{1/2}\,(1 - T/T_c)^{-1/2}$ and $\kappa \sim \lambda(0)/l$, it can be rewritten as $\kappa^2\,(\xi_0/l) \gg 1 - T/T_c$. If $\kappa \sim 1$, we find once again that it is less strict than the general condition $T_c - T \ll T_c$.

Thus, in the case of dirty superconductors (alloys), the GL theory is valid within a rather wide temperature interval – at least, qualitatively – and, provided the condition $T_c - T \ll T_c$ is satisfied, also quantitatively.

Problems

Problem 6.1. By how many times does the density of states in the elementary excitation spectrum of the superconductor exceed the density of electron states in the normal metal, in the vicinity of the Fermi surface? The values of the excitation energy E are $1.01\,\Delta_0$, $1.5\,\Delta_0$, and $2.0\,\Delta_0$.

Problem 6.2. Find the coupling constant $g = N(0)\,V$ for Sn, if $T_c = 3.74$ K and the Debye temperature is $\Theta_D = 195$ K.

Problem 6.3. The critical temperature of a mixture of Hg isotopes, having the average atomic weight 199.7 g, is 4.161 K. By how much will the critical temperature of the mixture change if the average atomic weight changes to 200.7 g? Will the temperature increase or decrease?

Problem 6.4. From a tunneling experiment (see Sect. 6.4), the energy gap of In was found to be $\Delta_0 = 5.3 \times 10^{-4}$ eV. What then is the critical temperature of In according to the BCS theory? Compare the result with the true value of T_c for In, 3.37 K.

Problem 6.5. Find the critical momentum $\hbar K_c$ of an electron pair if $v_F = 0.65 \times 10^8$ cm s^{-1} and $\Delta_0 = 0.58 \times 10^{-3}$ eV (these are the parameters corresponding to superconducting Sn). Compare K_c with k_F.

7. Some Nonequilibrium Effects in Superconductors

The question of what happens to a superconductor when its equilibrium is disturbed and how it relaxes back to the equilibrium state has been a subject of intense discussions. Why is this? Why are nonequilibrium properties of superconductors so interesting and important? One would think that the answer lies in the fact that the superconductor comprises several groups of electrons: superconducting electrons, electron-like excitations, and holes. What was elucidated following the publication of the BCS theory was the equilibrium situation, in which all electron groups are in equilibrium not only with each other but also with the crystal lattice. If the equilibrium is disturbed, the behavior of the system becomes more complex. Then, through analyzing the processes of relaxation back to equilibrium, we can learn more about the very basics of the theory and gain insight into the nature of the phenomena. In addition, one should bear in mind that many successful cryoelectronic devices normally operate under conditions which are far from equilibrium. This attaches to studies of nonequilibrium phenomena a practical importance, too.

In this chapter we shall discuss a number of cases in which a nonequilibrium state of a superconductor plays a decisive role in determining its physical properties.

7.1 Quasiparticles: Electrons and Holes

Elementary excitations in superconductors (electrons and holes) already were the subject of our discussion in Chap. 6. Now we shall define them more accurately.

Consider first the normal metal. Its ground state (at $T = 0$, without elementary excitations) is such that all free electrons occupy all the states inside the Fermi surface while all states outside it are empty. Let us now add one extra electron to the metal. It will occupy one of the states, $\boldsymbol{k}$, in momentum space ($k > k_F$, where k_F is the radius of the Fermi sphere), and the total energy of the electron system will increase by $\hbar^2(k^2 - {k_F}^2)/2m$. The energy, as before, is measured from the Fermi energy $\hbar^2 {k_F}^2/2m$. The increase in total energy will therefore be equal to the energy of the elementary excitation and the system's momentum will be $\boldsymbol{k}$, the momentum of the elementary excitation.

Consider once again the ground state of the normal metal, but this time after one electron has been removed from the state $\boldsymbol{k}$ ($k < k_F$). This state can also be treated as an elementary excitation of the system, because its energy is greater than the ground-state energy of electrons remaining in the metal, by $\hbar^2(k_F{}^2 - k^2)/2m$. Indeed, we can restore the ground state by taking an electron from the Fermi surface and putting it in the vacant state $\boldsymbol{k}$. The momentum of this elementary excitation is equal to the momentum of the entire electron system, that is, to $-\boldsymbol{k}$. Such an excitation (quasiparticle) behaves in the same way as a positive charge and is called a hole. The elementary excitation energy spectrum for the normal metal is shown in Fig. 7.1 by dashed lines.

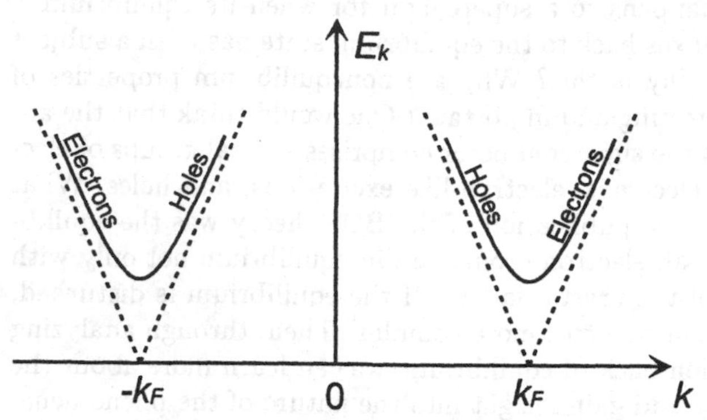

Fig. 7.1. Energy spectrum of the quasiparticles in the superconductor. Dashed lines show the spectrum for the normal metal

In the superconductor, things are somewhat more complicated. We already know (Chap. 6) that the ground state of the superconductor is described by a function $v_{\boldsymbol{k}}{}^2$ representing the probability that the pair states $(\boldsymbol{k}, -\boldsymbol{k})$ are occupied by electrons. The behavior of this function is illustrated in Fig. 6.5. Let us now add one extra electron to a superconductor. It will occupy one of the vacant $\boldsymbol{k}$ states in momentum space, while the state $-\boldsymbol{k}$ will remain empty. But, unlike the normal metal, in the superconductor, the state $\boldsymbol{k}$ can be either inside or outside the Fermi surface. Because some of the $\boldsymbol{k}$ states with $k < k_F$ remain empty in the ground state, too! Thus, in the superconductor, the momentum of an electron-like quasiparticle can be not only greater than, but also less than, k_F.

Conversely, if we remove one electron from, say, the state $-\boldsymbol{k}$ in the ground state of a superconductor, the complementary electron in the state $\boldsymbol{k}$ will become uncoupled and start behaving in the same manner as a hole with momentum $\boldsymbol{k}$ which, again, can be not only greater than, but also less than, k_F.

Therefore, in the superconductor, an elementary excitation corresponding to the state $\boldsymbol{k}$ being occupied and the state $-\boldsymbol{k}$ being empty behaves partly as an electron and partly as a hole. Its actual behavior is determined by the relative probabilities of the two states. The probability of its behaving as a hole must be proportional to the probability of the state $\boldsymbol{k}$ being occupied by a superconducting electron, that is, to $v_{\boldsymbol{k}}{}^2$. On the other hand, its behavior as an electron-like quasiparticle is governed by the probability that one of the $\boldsymbol{k}$ states in the ground state is vacant, that is, by $u_{\boldsymbol{k}}{}^2 = 1 - v_{\boldsymbol{k}}{}^2$.

One may define a quasiparticle of electron type in the superconductor (an almost pure electron) as a quasiparticle with momentum $\boldsymbol{k}$ satisfying the conditions $|k - k_{\mathrm{F}}|/k_{\mathrm{F}} \gg \Delta/\varepsilon_{\mathrm{F}}$ and $k > k_{\mathrm{F}}$. And an almost pure hole is an excitation with momentum $\boldsymbol{k}$ such that it satisfies the inequalities $|k - k_{\mathrm{F}}|/k_{\mathrm{F}} \gg \Delta/\varepsilon_{\mathrm{F}}$ and $k < k_{\mathrm{F}}$.

The electron and hole branches of the quasiparticle spectrum for the superconductor are sketched in Fig. 7.1 by solid lines.

It only remains to add that, in real space, the propagation velocity of a quasiparticle is the group velocity well known to be

$$s_{\boldsymbol{k}} = \frac{\mathrm{d}E_{\boldsymbol{k}}}{\mathrm{d}\boldsymbol{p}_{\boldsymbol{k}}} = \frac{1}{\hbar}\frac{\mathrm{d}E_{\boldsymbol{k}}}{\mathrm{d}\boldsymbol{k}} .$$

This means that at $k > 0$ (in the notation of Fig. 7.1) the holes move to the left and the electrons to the right.

7.2 Charge of a Quasiparticle in the Superconductor

We have just seen that a quasiparticle in the superconductor is an elementary excitation which can be treated to a certain extent as an electron and to a certain extent as a hole. As one moves along the curve of the elementary excitation spectrum, as in Fig. 7.1, the transition from the pure electron to the pure hole happens gradually. This simple observation suggests that one should assign a fractional electric charge to the quasiparticle. In the argumentation below we shall follow Pethick and Smith [67] who were apparently the first to propose this approach. We shall find very soon how convenient it is and how much easier it makes a qualitative analysis of nonequilibrium processes in superconductors. For the sake of simplicity, we agree to measure all charges in normalized units, taking the electron charge equal to unity.

Consider a superconductor at a finite temperature and assume that it contains both coupled (superconducting) electrons and elementary excitations. Suppose that the distribution of the elementary excitations in $\boldsymbol{k}$ space is defined by a distribution function $f_{\boldsymbol{k}}$, that is, $f_{\boldsymbol{k}}$ gives the probability that the state $\boldsymbol{k}$ is occupied by an excited electron. Since the probability that this state was free of a superconducting electron before is $u_{\boldsymbol{k}}{}^2$, the net probability to find an excited electron in the state $\boldsymbol{k}$ is given by $f_{\boldsymbol{k}} u_{\boldsymbol{k}}{}^2$. On the other hand, this state can be filled with one of the superconducting electrons. For this, two conditions must be satisfied simultaneously: first, that this state does not contain an excited electron (the probability of this is $1 - f_{\boldsymbol{k}}$), and second, that it is occupied by a superconducting electron (the probability equals v_k^2). Thus, the net probability to find a superconducting electron in the state $\boldsymbol{k}$ is $(1-f_k)\,v_k^2$ and the total charge associated with this state is $f_k u_k^2 + (1-f_k)\,v_k^2$. The total charge associated with all electrons in the superconductor is then

$$Q_{\text{tot}} = \sum_k f_k u_k^2 + \sum_k (1-f_k)\,v_k^2 \,. \tag{7.1}$$

Suppose that, as a result of an external perturbation, the distributions of both the quasiparticles and the superconducting electrons have changed (the former by the value of δf_k and the latter by the value of δv_k^2). Such a change will result in a variation of the total charge of the electron system by

$$\delta Q_{\text{tot}} = \sum_k (u_k^2 - v_k^2)\,\delta f_k + \sum_k (1-2f_k)\,\delta v_k^2 \,. \tag{7.2}$$

Here the first term is obviously a variation of the total charge of the quasiparticles. More specifically, the probability of finding a quasiparticle in the state $\boldsymbol{k}$ has changed by the value of δf_k and been accompanied by a change of the charge associated with this state by $(u_k^2 - v_k^2)\,\delta f_k$. It is obvious then that the quantity

$$q_k = u_k^2 - v_k^2 \tag{7.3}$$

can be treated as the charge of a quasiparticle with momentum $\boldsymbol{k}$. According to (6.12),

$$v_k^2 = \frac{1}{2}\left(1 - \frac{\varepsilon_k}{E_k}\right) . \tag{7.4}$$

Then

$$u_k^2 = \frac{1}{2}\left(1 + \frac{\varepsilon_k}{E_k}\right) . \tag{7.5}$$

This leads to

$$q_k = \frac{\varepsilon_k}{E_k} = \frac{\varepsilon_k}{(\varepsilon_k^2 + \Delta^2)^{1/2}} \,. \tag{7.6}$$

The dependence of q_k on k is shown in Fig. 7.2. It implies that the charge of a quasiparticle varies continuously from $+1$ (pure electron) to -1 (pure hole) and is completely determined by which particular state in $\boldsymbol{k}$ space the quasiparticle occupies. A rather odd result at first sight! But let us add one extra electron to the superconductor and place it in the state $\boldsymbol{k}$.

As a result, the superconductor as a whole has acquired a charge $+1$ (recall that it is this value that we agreed to regard as the electron charge), while the state $\boldsymbol{k}$ has got a charge q_k. But where is the difference $1-q_k$? We know that the total charge must be conserved! The answer is that the charge $1-q_k$ has been transferred to the superconducting electrons thereby increasing their charge. The whole process can be described as follows. An electron is added to the superconductor. It forms a couple with another electron from the state

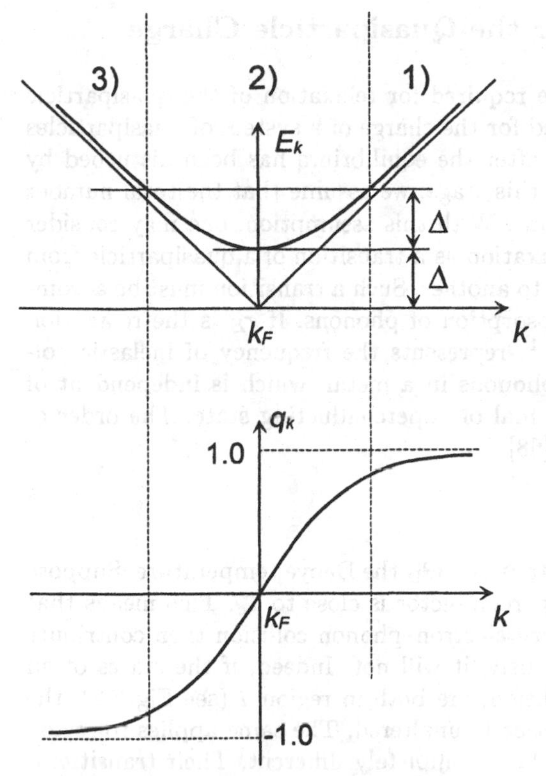

Fig. 7.2. Energy spectrum E_k of quasiparticles in the superconductor and their charge q_k. In region *1*, the quasiparticle charge q_k is close to +1, in region *3* it is close to −1, and in region *2* it may assume any value from +1 to −1

$-\boldsymbol{k}$ which belongs to the condensate[1] and the pair subsequently joins the condensate. But an electron in the state $\boldsymbol{k}$ is left without a partner, that is, a hole is formed.

Thus the condensate and the elementary excitations in the superconductor are closely interrelated and cannot be treated as independent in the relaxation process. Quasiparticle transitions between different $\boldsymbol{k}$ states are accompanied by changes of their charges q_k resulting in a change of the total charge of the condensate or, in other words, of the number of electrons in the condensate.

[1] We use the name 'condensate' for the entire sea of superconducting electrons, all of them described by a single wavefunction. It is, however, not entirely correct. Not all electron pairs, despite being Bose particles, can make a transition to the lowest energy level (i.e., to join the Bose condensate, in the strict meaning of the word). Owing to their interaction with each other, some of the Bose particles have energies above that of the condensate. Together with the condensate particles, they form the sea of superconducting electrons. On no account can these 'supercondensate' particles be regarded as elementary excitations.

7.3 Relaxation Time for the Quasiparticle Charge

Let us try to estimate the time required for relaxation of the quasiparticle charge, that is, the time required for the charge of a system of quasiparticles to attain its equilibrium value, after the equilibrium has been disturbed by some external perturbation. At this stage, we assume that the total number of quasiparticles remains constant. With this assumption, one may consider an elementary act of charge relaxation as a transition of a quasiparticle from one $\boldsymbol{k}$ state in momentum space to another. Such a transition must be accompanied by either emission or absorption of phonons. If τ_E is the relaxation time, the inverse quantity, ${\tau_E}^{-1}$, represents the frequency of inelastic collisions between electrons and phonons in a metal, which is independent of whether the metal is in the normal or superconducting state. The order of magnitude of this frequency is [68]

$$ {\tau_E}^{-1} \sim \frac{k_B T}{\hbar}\left(\frac{T}{\Theta_D}\right)^2 , $$

where k_B is the Boltzmann constant and Θ_D the Debye temperature. Suppose that the temperature of the superconductor is close to T_c. This means that $\Delta \ll k_B T_c$. Let us ask: Will every electron–phonon collision then contribute to the charge relaxation? Obviously, it will not. Indeed, if the states of an electron, before and after a collision, are both in region *1* (see Fig. 7.2), the transition leaves its charge practically unaltered. The same applies to region *3*. But the electrons in region *2* are completely different. Their transitions, initiated by collisions with phonons, alter the quasiparticle charge by values of the order of unity. At a finite temperature T, the fraction of excitations in region *2* out of the total number of excitations is of the order $\Delta / k_B T$ since, by definition, the excitation energies $E_{\boldsymbol{k}}$ in this region fall into the range $\Delta \le E_{\boldsymbol{k}} \le 2\Delta$. The frequency of transitions resulting in a considerable change of the quasiparticle charge is therefore

$$ {\tau_Q}^{-1} \sim \frac{\Delta}{k_B T} {\tau_E}^{-1} . $$

At $T \approx T_c$, rigorous calculations of the quasiparticle charge relaxation time [69] yield

$$ \tau_Q = \frac{4}{\pi} \frac{k_B T_c}{\Delta} \tau_E . \tag{7.7} $$

According to (7.7), the quasiparticle charge relaxation time diverges at $T \to T_c$. Typical values of τ_E are of the order 10^{-10} s. For example, for superconducting tin, $\tau_E = 3 \times 10^{-10}$ s; for lead $\tau_E = 10^{-11}$ s. The above values of τ_E correspond to the respective critical temperatures of the materials. The relaxation time τ_Q can be much longer. It depends entirely on how close one approaches the critical temperature.

7.4 Andreev Reflection

Let us examine the process in which the normal current converts into superconducting current. We shall try to understand what happens at the normal metal–superconductor interface and how far from the interface the conversion of the normal current into supercurrent takes place.

Before beginning our analysis, let us agree on what the term 'interface' means exactly. In its most pure form, it is the interface between a normal and a superconducting domain in the intermediate state of a superconductor. In this case, the order parameter (energy gap) has its maximum value $\Delta(T)$ deep inside the superconducting domain and is zero deep inside the normal region. The transition from one value to the other occurs over the coherence length $\xi(T)$. It is this transitional region to which we shall refer as the NS interface. In the case of mechanical contact between two metals – normal and superconducting – the situation remains the same, at least qualitatively. At $T \approx T_c$, the width of the interface, $\xi(T)$, is rather large and we can apply to it the following analysis. We are interested in what happens to an electron in the normal metal when, while moving towards the superconductor, it encounters the NS interface. We assume that the electron's energy is less than the energy gap of the superconductor. This process is illustrated in Fig. 7.3. At some moment of time while approaching the superconducting region, the normal electron reaches a point x which already belongs to the superconducting region but where the value of the gap $\Delta(x)$ is still small. At this moment it converts into an electron-like quasiparticle of the superconductor and, in momentum space, fills an appropriate $\boldsymbol{k}$ state, in accordance with the value of its energy $E_{\boldsymbol{k}}$. At the next moment, as the quasiparticle moves closer to the superconducting region, it reaches the area with a larger value of the gap thereby moving, in momentum space, to another $\boldsymbol{k}$ state, closer to k_F. This results in an increase of the quasiparticle charge.

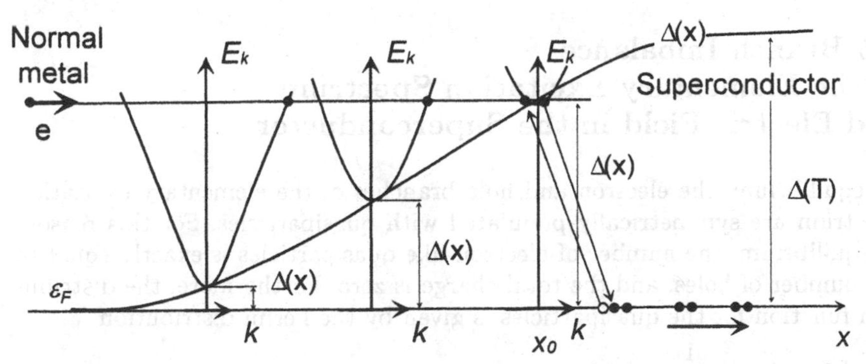

Fig. 7.3. Schematic diagram of Andreev reflection. Once in the vicinity of an NS interface, a quasiparticle (electron) moves from one $\boldsymbol{k}$ state to another, gradually changing its charge as it moves closer to the superconducting region

Thus, as an electron-like excitation moves from the normal metal towards the NS interface, it gradually reduces its charge. Finally, as one can see from the sketch in Fig. 7.3, the momentum of the quasiparticle becomes k_F, its group velocity zero and, according to (7.6), its charge reduces to zero ($q_{k_F} = 0$). This happens when the quasiparticle reaches the point x_0, where its energy is equal to the energy gap. At this point, the quasiparticle is reflected back from the interface and, in momentum space, moves to the left-hand branch of the energy spectrum, which is populated by holes. Now its group velocity is directed to the left, that is, in the direction from the superconductor towards the normal metal, and its charge becomes negative. But propagation of a negative charge to the left is equivalent to propagation of a positive charge to the right. Therefore, the process of reflection we have just analyzed gives rise to a charge transfer from the normal metal to the superconductor, i.e., to an electric current.

This process was first proposed theoretically by Andreev [70] and is called Andreev reflection.

The decrease of the quasiparticle charge, as it approaches the superconducting region, gives a clear indication that the condensate is also involved in the process of Andreev reflection. The quasiparticle charge does not disappear, it is transferred to the condensate! Physically, this means that, having reached the interface, the quasiparticle finds itself a partner and together they enter the condensate, while the newly formed hole goes back to the normal metal. This process accounts for the transfer through the NS interface of the part of the current that is carried by excitations with energies below the energy gap $\Delta(T)$. The process of normal current conversion into the supercurrent takes place within the region $\sim \xi(T)$. The behavior of excitations with energies $E_k > \Delta(T)$ is entirely different. As we are soon to find out, they can penetrate the interior of the superconductor to distances well in excess of the coherence length $\xi(T)$.

7.5 Branch Imbalance of the Elementary Excitation Spectrum and Electric Field in the Superconductor

At equilibrium, the electron and hole branches of the elementary excitation spectrum are symmetrically populated with quasiparticles. For this reason, at equilibrium, the number of electron-like quasiparticles is exactly equal to the number of holes, and the total charge is zero. Furthermore, the distribution function for the quasiparticles is given by the Fermi distribution

$$f_k = \frac{1}{\exp(E_k/k_B T) + 1} .$$

The equilibrium distribution can be disturbed as a result of an external perturbation. If, for example, a superconductor is exposed to electromagnetic radiation, the latter does not affect the symmetry of the distribution of electrons and holes over the branches of the spectrum. What happens is that the quasiparticles, having absorbed the electromagnetic energy, move up to higher energy levels, which may result in an alteration of the energy gap. Nonequilibrium processes of this type will be considered in Sect. 7.8.

Another way to disturb the equilibrium in the neighborhood of the NS interface is to have a flow of electrons coming from the normal metal. Indeed, in this process, electrons with energies $E_k > \Delta(T)$ penetrate the interior of the superconductor and occupy some of the $\boldsymbol{k}$ states on the electron branch of the spectrum. This causes a population imbalance on the two branches. In this section we want to investigate the consequences of this branch imbalance and the characteristic distances for its relaxation.

It is obvious that the existence of the branch imbalance leads to the appearance of a finite quasiparticle charge Q at a given location in the superconductor:

$$Q = \sum_k q_k f_k ,$$

where f_k is some (in the general case, nonequilibrium) distribution function for the electrons in momentum space.

But the net charge of all electrons at every point of the superconductor must be identical and equal in absolute value to the local charge on the ions of the crystal lattice. This condition assures electroneutrality of the material. Thus, an increase (compared to equilibrium) of the quasiparticle charge at some location by an amount Q must be compensated by a decrease, by the same amount, of the charge of the condensate, thereby modifying the distribution function of the superconducting electrons, $v_k{}^2$. The dotted line in Fig. 7.4 indicates the initial distribution function $v_k{}^2$ corresponding to equilibrium, when $Q = 0$ and the chemical potential of the superconducting electrons, μ_s, coincides with the Fermi energy ε_F of the metal.[2] Since the area under the $v_k{}^2$ curve is proportional to the number of superconducting electrons, a decrease of the latter ought to shift the distribution boundary to the left. This will result in a decrease of the chemical potential μ_s by the amount $\varepsilon_F - \mu_s$. The new distribution $v_k{}^2$ is shown in Fig. 7.4 by the solid line. Now we can easily count the number of states that have become free of superconducting electrons: $2\,N(0)\,(\varepsilon_F - \mu_s)$, where $N(0)$ is the density of states at the Fermi level, and the factor 2 takes into account that there are

[2] In the nonequilibrium case, the form of the electron distribution function for the condensate is the same as at equilibrium, i.e., $v_k{}^2 = (1 - \varepsilon_k/E_k)/2$, but with $\varepsilon_k = \hbar^2 k^2/2m - \mu_s$ and $E_k = (\varepsilon_k{}^2 + \Delta^2)^{1/2}$. That is, we assume that there is enough time for the condensate to adjust to a given nonequilibrium distribution of the quasiparticles. One can easily verify that the branch imbalance does not affect the value of the energy gap.

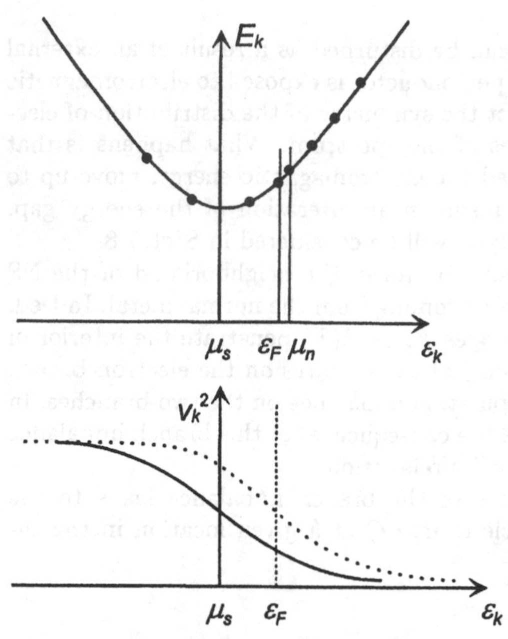

Fig. 7.4. Nonequilibrium distribution of superconducting electrons, $v_k{}^2$, and quasiparticle spectrum E_k for the superconductor in a nonequilibrium state. The dotted line shows the equilibrium distribution $v_k{}^2$ for $\mu_s = \varepsilon_F$. In the latter case, the energy gap Δ retains its value while the chemical potential μ_n becomes greater than ε_F

two electrons with opposite spins at each energy level. It follows that the condensate has lost the charge

$$Q = 2\,N(0)\,(\varepsilon_F - \mu_s)\ , \tag{7.8}$$

which has subsequently been added to the quasiparticles.

Let us analyze the consequences of this process. Since the nonequilibrium quasiparticle charge Q becomes a function of the coordinate x, it is obvious from (7.8) that, in real space, the nonequilibrium chemical potential of the superconducting electrons, μ_s, also becomes a function of x. But this is equivalent to the possibility of having an electric field $\boldsymbol{E}$ in the superconductor, such that it does not accelerate the condensate. Indeed, the most general form of the equation of motion for a superconducting electron reads

$$\frac{\mathrm{d}}{\mathrm{d}t}(m\boldsymbol{v}_s) = -\frac{e}{c}\frac{\mathrm{d}\boldsymbol{A}}{\mathrm{d}t} - e\nabla\varphi - \nabla\mu_s\ . \tag{7.9}$$

Here e is the electron charge, m the electron mass, $\boldsymbol{v}_s$ the superfluid velocity of the condensate, φ the electrostatic potential, and $\boldsymbol{A}$ the vector potential. Under stationary conditions, $\mathrm{d}(m\boldsymbol{v}_s)/\mathrm{d}t = \mathrm{d}\boldsymbol{A}/\mathrm{d}t = 0$, and the superconducting electrons do not accelerate. This implies that the electrochemical potential of the superconducting electrons, $\varphi_{\mathrm{ech,\,s}}$, must be constant in space:

$$\varphi_{\mathrm{ech,\,s}} = e\varphi + \mu_s = \mathrm{const}\ ; \tag{7.10}$$

however, it does not oppose the existence of the electric field. The latter is, from (7.10),

$$\boldsymbol{E} = -\nabla\varphi = \frac{1}{e}\nabla\mu_{\mathrm{s}} .$$

Using (7.8) we have

$$\boldsymbol{E} = -\frac{1}{2e\,N(0)}\nabla Q . \tag{7.11}$$

The above analysis infers that a gradient of the quasiparticle charge in a superconductor under nonequilibrium conditions brings about an electric field. However, this field does not cause the condensate to accelerate, because it is compensated by a gradient of the chemical potential of superconducting electrons. We emphasize once again that superconducting electrons are only affected by the gradient of their electrochemical potential $\nabla\varphi_{\mathrm{ech,s}}$ which, by (7.10), is zero.

For a given site in the superconductor, the presence of a stationary but nonequilibrium quasiparticle charge of density Q indicates that there exists a continuous flow of quasiparticles through this site. The quasiparticles bring with them a certain charge which is immediately transferred to the condensate through the relaxation process.

The entire process can be expressed by a simple relation

$$\operatorname{div}\boldsymbol{j}_{\mathrm{n}} = -eQ/\tau_Q , \tag{7.12}$$

where $\boldsymbol{j}_{\mathrm{n}}$ is the normal component of the total current. The relation (7.12) is only approximately valid when the gap is small ($\Delta \ll k_{\mathrm{B}}T_{\mathrm{c}}$), that is, when the temperature is close to T_{c}. One can find a more rigorous treatment of this process in [67]. In the same approximation, at $T \approx T_{\mathrm{c}}$, the change in the chemical potential of the quasiparticles is small compared to that of the superconducting electrons, and can be neglected, i.e., $\varepsilon_{\mathrm{F}} - \mu_{\mathrm{s}} \gg \mu_{\mathrm{n}} - \varepsilon_{\mathrm{F}}$ (see Fig. 7.4). Indeed, at $T \approx T_{\mathrm{c}}$, the number of superconducting electrons in the condensate is small, while that of normal electrons is substantial. Therefore, the removal of a few electrons from the condensate and their transfer to the sea of normal electrons will result in a considerable alteration of the properties of the condensate (a decrease in μ_{s}) while leaving the normal electrons almost unaffected.

In view of this, one may write Ohm's law for the excitations in the usual form

$$\boldsymbol{j}_{\mathrm{n}} = \sigma\boldsymbol{E} ,$$

where σ is the normal conductivity at low temperatures: $\sigma^{-1} = \rho_{\mathrm{res}}$, and ρ_{res} the residual resistivity.[3] Taking the divergence of both sides of this equation and substituting (7.11) and (7.12) yields

[3] The residual resistivity of a conductor is the resistivity extrapolated to $T = 0$.

$$\nabla^2 Q = \frac{1}{\lambda_Q{}^2} Q \,, \tag{7.13}$$

where

$$\lambda_Q{}^2 = \frac{\sigma \, \tau_Q}{2 \, e^2 \, N(0)} \,. \tag{7.14}$$

Taking into account that the conductivity of the normal metal can be written as[4]

$$\sigma = \frac{2}{3} e^2 \, N(0) \, l v_{\mathrm{F}} \,,$$

where l is the electron mean free path and v_{F} is the electron velocity on the Fermi surface, we finally have

$$\lambda_Q = \left(\frac{l v_{\mathrm{F}} \tau_Q}{3} \right)^{1/2} = (D \tau_Q)^{1/2} \,, \tag{7.15}$$

where D is the electron diffusion coefficient, $D = l v_{\mathrm{F}}/3$.

Consider now the simplest case. Suppose that the sample is a filament so thin that the problem can be considered one-dimensional. Let us take the x axis along the filament and assume the material of the filament to be normal for $x < 0$ and superconducting for $x > 0$. Further, suppose that the filament carries a current in the positive x direction. Equation (7.13) then takes the form

$$\frac{\mathrm{d}^2 Q}{\mathrm{d}x^2} = \frac{1}{\lambda_Q{}^2} Q \,. \tag{7.16}$$

It is obvious that far away from the interface (at $x \to \infty$) the charge is $Q \to 0$, because there the superconductor is at equilibrium and the quasiparticle charge is zero. The solution of (7.16) is then given by the function

$$Q \propto \mathrm{e}^{-x/\lambda_Q} \,. \tag{7.17}$$

Accordingly, the charge gradient gives rise to an electric field in the interior of the superconductor which, by (7.11), will be

$$E = E_0 \, \mathrm{e}^{-x/\lambda_Q} \,, \tag{7.18}$$

where E_0 is the electric field at the interface. If one ignores the Andreev reflection, which is indeed permitted at $T \approx T_{\mathrm{c}}$, the field E_0 at the interface will be equal to the electric field in the interior of the normal metal, i.e., at $x < 0$.

Thus the quantity λ_Q represents the depth to which the electric field penetrates into the superconductor. According to (7.18), the electric field is exponentially damped over a distance λ_Q from the interface. Let us estimate the electric field penetration depth. According to (7.14), $\lambda_Q \propto \tau_Q{}^{1/2}$ and, according to (7.7), τ_Q is proportional to Δ^{-1}. Taking into account that $\Delta \propto (1 - T/T_{\mathrm{c}})^{1/2}$, we have

[4] See, for example, [4].

$$\lambda_Q \propto \left(1 - \frac{T}{T_c}\right)^{-1/4} , \tag{7.19}$$

that is, the depth λ_Q diverges at $T \to T_c$. Thus, at temperatures sufficiently close to the critical temperature, the electric field can penetrate the superconductor to large distances. Recalling that the characteristic value of τ_Q is of the order 10^{-9}–10^{-10} s, we obtain for $v_F \sim 10^8$ cm s^{-1} and, say, $l \sim 10^{-5}$ cm: $\lambda_Q \sim 10^{-4}$–10^{-3} cm. This means that the electric field penetration depth can surpass all other characteristic lengths of the superconductor. Not too close to T_c, it is possible to have $\lambda_Q \gg \xi(T)$.

Let us summarize. An electric current passing through the interface between a superconductor and a normal metal gives rise to an electric field in the superconductor. The field decays over the distance λ_Q from the interface, where λ_Q can be of a macroscopically large value. Within this region, the divergence of the normal current, $\nabla \boldsymbol{j}_n$, is nonzero (see (7.12)). But for the total current $\boldsymbol{j} = \boldsymbol{j}_s + \boldsymbol{j}_n$ (where $\boldsymbol{j}_s$ is the supercurrent density) under stationary conditions, the divergence is, naturally, zero. Therefore, we have

$$\nabla \boldsymbol{j}_n = -\nabla \boldsymbol{j}_s .$$

It means that within the layer of a superconductor of thickness λ_Q near the interface with the normal metal, the normal current is converted into the supercurrent. This can also be interpreted in a different way. According to Fig. 7.4, the chemical potentials in this 'nonequilibrium' region are: $\mu_n > \mu_s$. Therefore, one may say that this region hosts a 'chemical reaction' in which a substance with chemical potential μ_n transforms into a substance with chemical potential μ_s.

All the processes analyzed in this section relate to the case where there is a current of electrons passing from a normal metal to a superconductor. However, it is easy to verify that the entire pattern is symmetric with respect to current reversal. In that case, the particles moving from the normal metal towards the superconductor and entering the interface are holes, and the Andreev-reflected particles are electron-like. Furthermore, the branch imbalance of the elementary excitations in the superconductor, in the region $\sim \lambda_Q$ near the interface, is such that the population of the hole branch exceeds that of the electron branch.

7.6 Experimental Study of a Nonequilibrium State at a Normal Metal–Superconductor Interface

7.6.1 The Experiments of Yu and Mercereau

In this section we shall describe particularly elegant experiments due to Yu and Mercereau [71]. These experiments have clearly demonstrated that, in the presence of an electric current passing through an NS interface, the electrochemical potentials of the normal and superconducting electrons indeed differ from each other within a region of the superconductor adjacent to the normal metal.

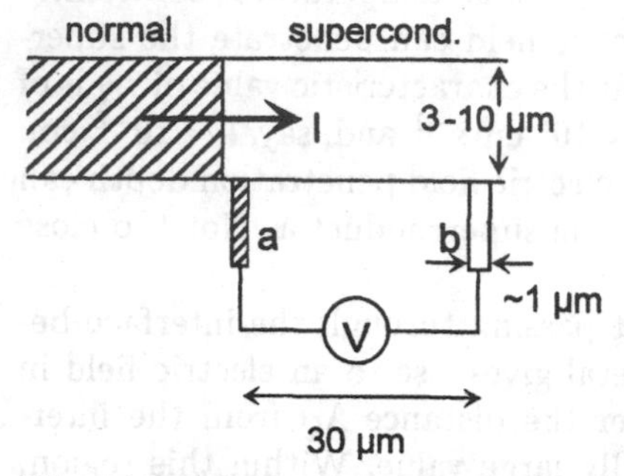

Fig. 7.5. Schematic diagram of the experiment by Yu and Mercereau [71]. Voltage leads a and b are connected to the voltmeter. If lead a is normal, the voltmeter registers the difference between electrochemical potentials of the quasiparticles and the condensate. If lead a is superconducting, the voltmeter reading is zero. Note that the readings do not depend on whether lead b is in the normal or superconducting state

The critical temperature of a thin film of tantalum of thickness of about 250 Å is approximately 4.1 K. If a part of the film is made thinner (about 100 Å) by anodization, the critical temperature of this part will fall to 3.5 K. Further, if the temperature of the film is kept between 3.5 K and 4.1 K, one part of the film will be normal and the other superconducting.

If we now apply a current to the film, the electrochemical potential of the superconducting electrons, $\varphi_{\mathrm{ech,s}}$, will be constant in space, according to (7.10). Therefore, the superconducting voltage leads a and b (see Fig. 7.5)

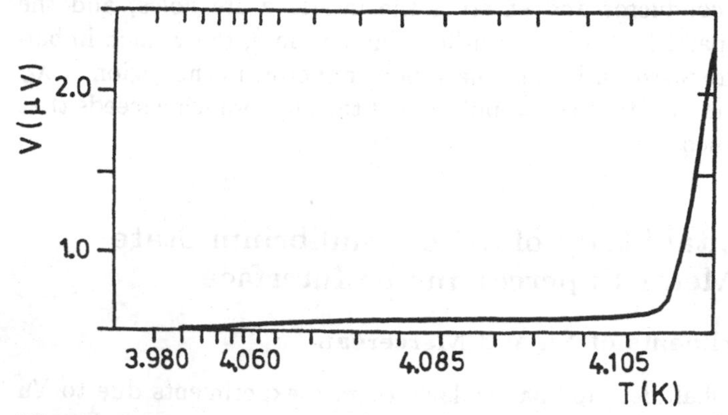

Fig. 7.6. Results of the Yu–Mercereau experiment [71]. The voltmeter registers the difference between the electrochemical potentials of the quasiparticles and the condensate in the nonequilibrium region

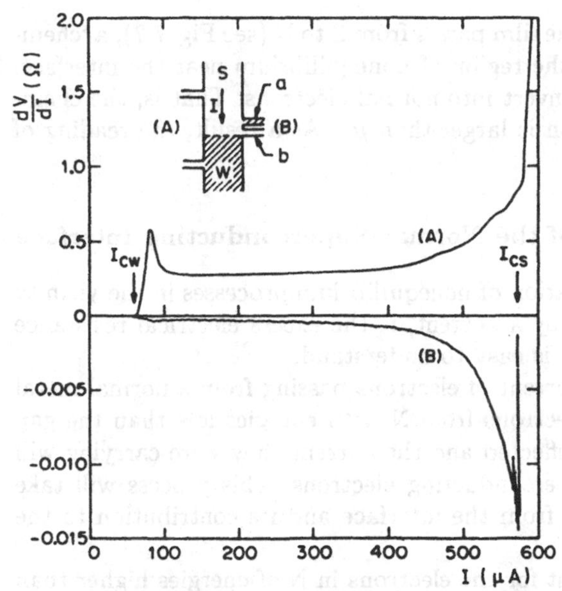

Fig. 7.7. Results of the Yu–Mercereau experiment [71]: differential resistance as a function of current through the film. Voltage A is opposite in sign to voltage B. Current flows in the direction from the superconducting to the normal (*dashed area*) lead

will convey to the voltmeter equal electrochemical potentials. That is, the voltmeter will register zero voltage. If, however, one of the voltage leads, a, immediately adjacent to the NS interface is made of a normal metal, it will convey to the left input contact of the voltmeter the electrochemical potential of the normal electrons, which in this region differs from $\varphi_{\text{ech, s}}$. This difference relaxes over the distance λ_Q from the interface. If the distance between the voltage leads a and b exceeds λ_Q, the lead b will send to the right contact of the voltmeter the electrochemical potential of the superconducting electrons (in equilibrium conditions, the latter is equal to the electrochemical potential of the quasiparticles). As a result, the voltmeter will measure the difference in electrochemical potential between the normal and superconducting electrons in the vicinity of the NS interface.

The result of the experiment with the lead a normal is shown in Fig. 7.6.

Another experiment by the same authors proved to be even more elegant. By placing the voltage contacts in the nonequilibrium region of a superconductor close to each other (as illustrated in Fig. 7.7) and in such a way that the superconducting lead b was very close to the interface and the normal lead a was some distance away from it, they were able to observe a reversal of the voltage sign for the same direction of current as before.

These experiments have provided ample verification of the ideas discussed in the foregoing sections.

Indeed, if the current in the film passes from S to N (see Fig. 7.7), a 'chemical reaction' takes place in the region of nonequilibrium near the interface: superconducting electrons convert into normal electrons. That is, the chemical potential μ_s in this region is larger than μ_n. As a result, the reading of the voltmeter changes sign.

7.6.2 Excess Resistance of the Normal–Superconducting Interface

Another interesting manifestation of nonequilibrium processes in the vicinity of an NS interface, initiated by a current, is the excess electrical resistance of the interface. Physically, it is easy to understand.

Suppose that there is a current of electrons passing from a normal metal N to a superconductor S. Electrons from N with energies less than the gap $\Delta(T)$ in S will be Andreev-reflected and the current they were carrying will convert into a current of superconducting electrons. This process will take place within a distance $\xi(T)$ from the interface and its contribution to the resistivity will be zero.

But things will be different for the electrons in N of energies higher than the gap $\Delta(T)$. Upon arriving in S, they move, in momentum space, to the electron branch of the elementary excitation spectrum of the superconductor. This, as was established in the previous section, results in an imbalance of the quasiparticle charge Q and gives rise to an electric field E which extends into the superconductor over a distance λ_Q from the interface. Furthermore, the presence of an electric field in the region λ_Q near the interface implies that this region contributes to the net voltage drop, that is, to the net resistance of the circuit. This additional resistance is called the excess resistance of an NS interface. If all electrons carrying the current were accelerated by the electric field, the excess resistance would simply be

$$R_{\text{exc}} = \rho\lambda_Q/S\,,$$

where ρ is the normal-state resistivity of the superconductor and S its cross-sectional area. One should bear in mind, however, that not all electrons are accelerated by the electric field E.

Indeed, as we already know, the electrons of the condensate are not accelerated by the field E, because the effect of the field on superconducting electrons is exactly compensated by the gradient of their chemical potential $\nabla\mu_s$. This property is a characteristic of all electrons passing from N into S with energies that are below the gap $\Delta(T)$ (they are Andreev-reflected), and also of the quasiparticles with energies above the gap: $E_k \geq \Delta(T)$. The number of quasiparticles which do not contribute to the excess resistance is determined by the following process. Consider an electron having a charge q_k which is injected from N into S. Upon entering S, the electron will have 'lost' a part of its charge $1-q_k$ which has been transferred to the condensate. The transmission of this part of the charge through the NS interface occurs via Andreev reflection. Therefore, the part $1-q_k$ does not contribute to the excess resistance. The latter should then be written as

$$R_{\text{exc}} = \frac{Z(T)\,\rho\lambda_Q}{S}\,,$$

where the coefficient $Z(T)$ accounts for the number of electrons which are *not* Andreev-reflected at the interface.

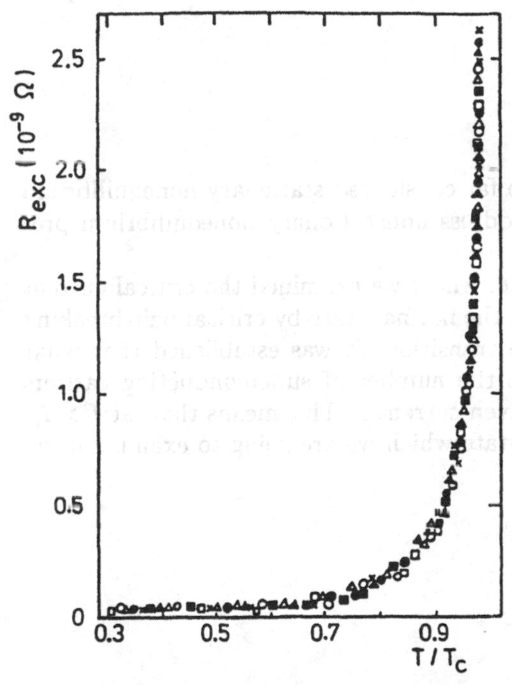

Fig. 7.8. Excess resistance of an NS interface as a function of temperature, measured on Ta-Cu-Ta sandwiches [72] The data were obtained from seven different samples

It is obvious that as $T \to 0$ all electrons passing through the interface are Andreev-reflected and $Z \to 0$. Conversely, as $T \to T_c$ we have $\Delta(T) \to 0$ so that the overwhelming majority of electrons pass through the interface without being Andreev-reflected and therefore contribute to the resistance. In this limit $Z \to 1$.

Therefore, in the entire temperature range, $R_{\text{exc}}(T)$ is a monotonic function of temperature:[5]

$$R_{\text{exc}}(T \to 0) = 0, \qquad R_{\text{exc}}(T \to T_c) = \rho\lambda_Q/S\,.$$

In the latter case ($T \to T_c$), the temperature dependence of R_{exc} is governed

[5] The function $Z(T)$ is calculated in a number of theoretical papers. See, for example, [73, 74].

by the temperature dependence of λ_Q (7.19):

$$\lambda_Q \propto \left(1 - \frac{T}{T_c}\right)^{-1/4} .$$

This conclusion agrees well with experiment.

As an illustration, the results of experiments on seven different SNS sandwiches are shown in Fig. 7.8. The excess resistances of NS interfaces are plotted as functions of temperature.

7.7 Phase-Slip Centers

In the present chapter, we have so far considered stationary nonequilibrium processes. Now we are going to address nonstationary nonequilibrium processes.

It was demonstrated in Sect. 3.6, where we examined the critical current of a thin film, that a film driven to the normal state by critical pair-breaking current does not undergo a phase transition. It was established that what happens at $I = I_c$ is simply that the number of superconducting carriers becomes insufficient to sustain a given current I. This means that, at $I > I_c$, there appears a peculiar resistive state which we are going to examine now.

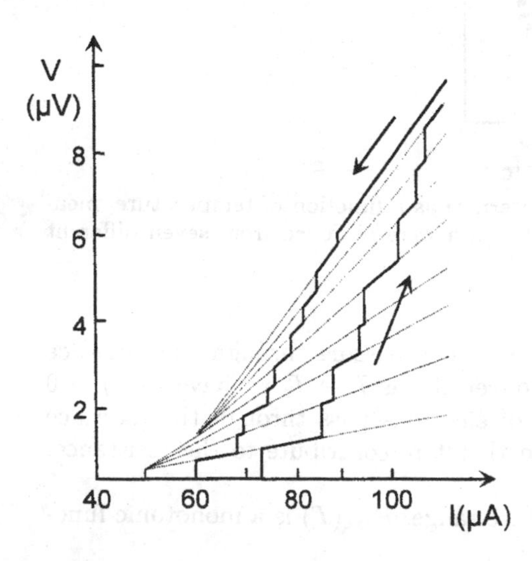

Fig. 7.9. Current–voltage characteristic of a single-crystalline Sn filament: the length of the filament is 0.8 mm and its cross-sectional area is 1.93 mm^2 [75]

Let us start with experimental results. The most suitable objects to study this kind of resistive state are quasi-one-dimensional objects, such as long thin superconducting filaments of diameter $d \leq \xi(T)$. The diameter of such a filament is so small that an Abrikosov vortex having a normal core (of radius

$\sim \xi(T)$) cannot fit into it. Therefore, the resistive state of the filament is not masked by effects typical of the resistive state of type-II superconductors. The current–voltage characteristic of a thin single-crystalline whisker of tin is shown in Fig. 7.9. One can see that the voltage builds up in a stepwise fashion and each upward jump of the voltage is accompanied by an increase of the sample resistivity (the slope of the I–V characteristic).

This behavior in the resistive state is explained by the model proposed by Skocpol, Beasley and Tinkham [76], the so-called phase slip center model.

Suppose that at a certain location (weak spot) in the filament the critical current is somewhat smaller than elsewhere. Then, as we increase the current through the filament, the current will first reach its critical value at the weak spot. A further increase of the current will result in a flow of normal electrons which will induce an electric field. The latter, in turn, will accelerate the normal electrons to the critical velocity. Electron pairs at the weak spot will be broken, the amplitude of the order parameter $|\psi|$ will go to zero and the current will be carried by the normal component alone. But formation of electron pairs remains favored. Therefore, the amplitude of the order parameter $|\psi|$ will become nonzero again and the current will again be carried partly by superconducting electrons, i.e., a supercurrent will appear. Then the entire process will repeat itself. On completion of each cycle, the phase difference of the superconducting electron wavefunctions to the left and to the right of the weak spot will vary by 2π; hence the name phase slip center. The size of the center is given, of course, by the size of the region where the oscillations of $|\psi|$ take place, which is $\sim 2\xi(T)$.

At the moment when $|\psi| = 0$, a phase slip center is in the normal state and the electric field penetrates the parts of the superconductor adjacent to it, to a depth $\sim \lambda_Q$ (see Sect. 7.5). Therefore, formation of one phase slip center is accompanied by the appearance of a finite resistance $\rho\, 2\lambda_Q/S$, where ρ is the normal-state resistivity of the material of the filament and S is its cross-sectional area. The voltage drop across this resistance is produced solely by the normal component of the current: $I_n = I - I_s$. Let us find the average of this expression over time. I is a given constant current independent of time. The supercurrent I_s oscillates between I_c and 0. Therefore, we assume $\overline{I}_s = \beta I_c$, where $\beta \sim 0.5$. Then, for the average voltage across one phase slip center we have

$$\overline{V} = \frac{2\lambda_Q \rho(1 - \beta I_c)}{S} . \tag{7.20}$$

Despite its simplicity, this formula describes the experimental results very well.

A further increase of the current will give rise to a second phase slip center, then a third and so on. The arrival of every new center will cause the total voltage across the filament to increase in a jump, and the slope of the current–voltage characteristic will increase, too, owing to the increased number of resistive regions. One can trace this behavior in Fig. 7.9.

7.8 Nonequilibrium Enhancement of Superconductivity

In this section we shall examine the case of a symmetric deviation from the equilibrium distribution function, such that the nonequilibrium distributions of the quasiparticles for the two branches of the elementary excitation spectrum are identical. Contrary to the other nonequilibrium processes which we have already analyzed, a deviation from equilibrium in this case brings about a change of the energy gap Δ.

A symmetric nonequilibrium distribution can be realized, for instance, by exposing a superconductor to electromagnetic radiation of frequency $\omega < 2\Delta/\hbar$. Then a quantum of electromagnetic energy incident on the superconductor will be insufficient to break a pair but will still be enough to move elementary excitations up to the higher energy levels of the excitation spectrum. Such a nonequilibrium distribution of quasiparticles ought to cause an increase of both the energy gap and the critical temperature.

Indeed, by making the lower energy levels of the elementary excitation spectrum vacant, we open the way for the electron pairs of the condensate to make transitions to these $\boldsymbol{k}$ states thereby increasing the number of terms in the sum of (6.34), which defines the energy gap. A number of experimental results support these ideas. For instance, an increase of the critical current of narrow superconducting constrictions was observed in [77], after the sample had been subjected to ultra-high-frequency radiation. Theoretical treatment of these effects can be found in works by Eliashberg [78].

7.9 Thermoelectric Effect in Superconductors

For a long time it was believed that there are no thermoelectric effects in superconductors. However, this is not true. To find out what really happens in superconductors, let us first consider the ordinary thermoelectric effect in a normal metal.

Suppose that the two ends of a bulk piece of a normal metal have different temperatures, T_1 and T_2, that is, there is a temperature gradient ∇T across the metal. Free electrons in the metal are acted upon by a force pushing them in the direction from the 'hot' to the 'cold' end. A subsequent accumulation of opposite charges at the opposite ends of the superconductor produces an electric field in its interior which, in turn, exerts a force on the electrons in the reverse direction, i.e., from the cold to the hot end. Under stationary conditions, the two forces balance each other resulting in a constant voltage drop across the specimen, V, referred to as the thermo-electromotive force (thermo-emf):

$$V = \alpha (T_1 - T_2) ,$$

where α is the absolute thermoelectric power (thermopower) of the metal.

In the superconductor, the process of heat transfer relates to entirely different electron phenomena.[6] Suppose, again, that two ends of a bulk piece of metal (this time of a superconductor) are at different temperatures, T_1 and T_2. Again, the temperature gradient ∇T will produce a force on normal excitations of the superconductor initiating a current of the normal excitations

$$\boldsymbol{j}_{\rm n} = \sigma\alpha\nabla T , \tag{7.21}$$

where σ is the normal-state conductivity of the superconductor. As an immediate response, a countercurrent of the superconducting component, $\boldsymbol{j}_{\rm s}$, will start, thereby compensating the normal current $\boldsymbol{j}_{\rm n}$. Indeed, according to (7.21), the current $\boldsymbol{j}_{\rm n}$ is a potential current, i.e., $\operatorname{curl} \boldsymbol{j}_{\rm n} = 0$. This means that for the total current $\boldsymbol{j}$ defined as

$$\boldsymbol{j} = \boldsymbol{j}_{\rm n} + \boldsymbol{j}_{\rm s} ,$$

we have $\operatorname{curl} \boldsymbol{j} = \operatorname{curl} \boldsymbol{j}_{\rm s}$. The supercurrent $\boldsymbol{j}_{\rm s}$ must, under stationary conditions, satisfy the London equation. Taking the curl of both sides of (2.17) we get

$$\operatorname{curl} \Lambda \boldsymbol{j}_{\rm s} = -\frac{1}{c}\boldsymbol{H} ,$$

which means that in our case the equation

$$\operatorname{curl} \Lambda \boldsymbol{j} = -\frac{1}{c}\boldsymbol{H} \tag{7.22}$$

is valid. Using Maxwell's equations $\operatorname{curl} \boldsymbol{H} = \frac{4\pi}{c}\boldsymbol{j}$ and $\operatorname{div} \boldsymbol{H} = 0$, we then arrive at the well-known equation for the magnetic field:

$$\nabla^2 \boldsymbol{H} = \frac{1}{\lambda^2}\boldsymbol{H} ,$$

from which the Meissner–Ochsenfeld effect follows directly. This implies that, in the presence of heat flow, the total current in the interior of a bulk isotropic homogeneous superconductor must be zero, too:

$$\boldsymbol{j} = \boldsymbol{j}_{\rm n} + \boldsymbol{j}_{\rm s} = 0 ,$$

that is, $\boldsymbol{j}_{\rm n} = -\boldsymbol{j}_{\rm s}$.

This means that, as they approach the edge of the superconductor, normal excitations driven by the temperature gradient convert into superconducting carriers, i.e., electron pairs, and move in the reverse direction, thereby forming the countercurrent $\boldsymbol{j}_{\rm s}$. A graphic illustration of this process is given by the equivalent circuit shown in Fig. 7.10.

Let us assume that the normal excitations are acted upon by an electromotive force (emf) E_T produced by the temperature gradient:

$$E_T = \alpha(T_1 - T_2) .$$

The emf induces a current $I_{\rm n} = E_T/R_{\rm n}$ in the circuit, which flows back

[6] See the reviews [79–81].

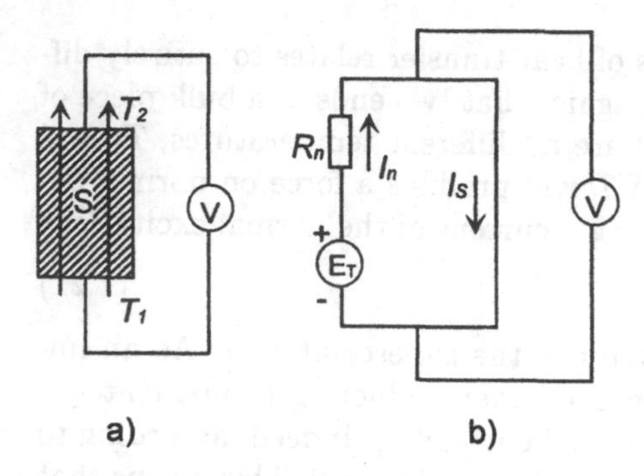

Fig. 7.10. (a) Superconductor carrying a heat flow, $T_1 > T_2$; (b) equivalent circuit for a superconductor subjected to a temperature gradient. R_n is the specimen resistance in the normal state, E_T is the thermo-emf acting upon the normal excitations

through the superconducting region as a supercurrent I_s. With this in mind, we can understand why the reading of the voltmeter incorporated in the circuit of Fig. 7.10 is zero: it is shunted by the supercurrent. That is why it was presumed for a long time that there are no thermoelectric effects in superconductors.

7.10 Superconducting 'Thermocouple' and the Magnetic Flux Induced by a Heat Flow

Now we shall see that thermoelectric effects can indeed be observed in superconductors. Consider two bulk specimens of two dissimilar superconductors, S_1 and S_2, which are brought into contact in such a way that together they form a closed ring, as in Fig. 7.11. We assume that the contact regions have different temperatures, T_1 and T_2.

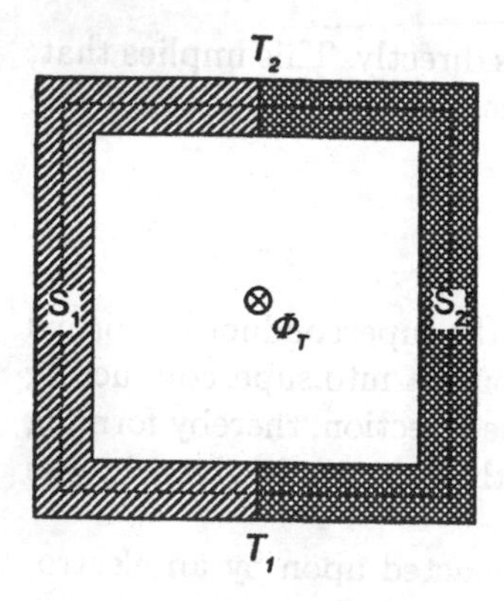

Fig. 7.11. Superconducting 'thermocouple'. The magnetic flux Φ_T through the ring is proportional to the temperature difference $(T_1 - T_2)$

The temperature gradient induces a current $\boldsymbol{j}_n = \sigma\alpha\nabla T$ in each half of the ring, S_1 and S_2. The supercurrent, which starts in response in the interior of the superconductor, is $\boldsymbol{j}_s = -\boldsymbol{j}_n$ and, according to the second Ginzburg–Landau equation, can be written as

$$j_s = \frac{\hbar n_s e}{2m} \nabla\theta - \frac{e^2}{mc} n_s \boldsymbol{A}, \tag{7.23}$$

where n_s is the superconducting electron density, e and m are the charge and mass of a free electron, respectively, θ is the phase of the superconducting wavefunction of the GL theory, and $\boldsymbol{A}$ is the vector potential.

Taking into account (7.21), we have

$$\nabla\theta = \frac{2m}{\hbar n_s e}(-\sigma\alpha\nabla T) + \frac{2e}{\hbar c}\boldsymbol{A}. \tag{7.24}$$

As a next step, we integrate this equation over the dashed contour which passes through the interior of our bimetallic superconducting ring in such a way that its distance from the surfaces of the ring everywhere exceeds the penetration depth λ (see Fig. 7.11). Taking into account the requirement of the GL theory that the wavefunction must be single-valued leads to an equation which is already well-known to us:

$$\oint \nabla\theta \, dl = 2\pi n, \quad n = 0, 1, 2, \dots . \tag{7.25}$$

Integrating the right-hand side of (7.24) and using (7.25) yields

$$2\pi n = -\frac{2m}{\hbar e}\int_{T_1}^{T_2}\left(\frac{\sigma_1\alpha_1}{n_{s1}} - \frac{\sigma_2\alpha_2}{n_{s2}}\right) dT + \frac{2\pi\Phi}{\Phi_0}.$$

This means that the total magnetic flux through the bimetallic ring can be expressed as

$$\Phi = \Phi_0 n + \Phi_T, \tag{7.26}$$

where

$$\Phi_T = \Phi_0 \frac{m}{\pi e\hbar}\int_{T_1}^{T_2}\left(\frac{\sigma_1\alpha_1}{n_{s1}} - \frac{\sigma_2\alpha_2}{n_{s2}}\right) dT. \tag{7.27}$$

The indices 1 and 2 in the integrand refer to S_1 and S_2, respectively.

It follows from (7.26) and (7.27) that even in the case $n = 0$, that is, when the magnetic flux through the ring is zero, the difference in temperature between the contact regions S_1 and S_2 induces a magnetic flux Φ_T. This flux is generated by the supercurrent circulating along the inner surface of the bimetallic ring. If the quantities σ, α and n_s are assumed to be independent of temperature in the interval $T_1 < T < T_2$, (7.27) can be simplified to

$$\Phi_T = \Phi_0 \frac{m}{\pi e\hbar}\left(\frac{\sigma_1\alpha_1}{n_{s1}} - \frac{\sigma_2\alpha_2}{n_{s2}}\right)\Delta T,$$

where $\Delta T = T_2 - T_1$.

Taking $\sigma \sim 10^7\ (\Omega\,\mathrm{m})^{-1}$, $\alpha \sim 10^{-6}\ \mathrm{V\,K^{-1}}$, and $n_s \sim 10^{28}\ \mathrm{m^{-3}}$ we obtain for $\Phi_T/\Delta T$ a value of the order $10^{-5}\,\Phi_0$ per Kelvin. Measuring such tiny variations of magnetic flux is near the limit of SQUID sensitivity.

At temperatures close to T_c, the effect is enhanced, since the superconducting electron density there tends to zero. A magnetic flux Φ_T induced by a temperature gradient in a bimetallic ring was indeed observed in the experiments reported in [82, 83]. Its value, however, proved to be much larger that the theoretical estimate. The cause of this discrepancy is not clear yet.

7.11 Thermoelectric Effects in Josephson Junctions

One encounters very interesting and striking thermoelectric effects in Josephson junctions. They are most pronounced in superconductor–normal metal–superconductor (SNS) junctions. As we are soon to find out, the reason for this lies in the fact that phase coherence in this type of Josephson junctions is established through a fairly thick normal layer ($\sim 10^{-3}$ cm). The typically very low resistance of the normal layer results in very small characteristic values of the voltage across the junction, V_c. In this section we shall first consider the behavior of an SNS junction when its S electrodes have different temperatures or, in other words, when there is a heat flow through the junction. After that, we shall discuss the thermoelectric effects arising when, in addition to carrying a heat flow, the junction is placed in a magnetic field.

7.11.1 SNS Josephson Junction in a Temperature Gradient

Let us consider an SNS Josephson junction placed in a temperature gradient (Fig. 7.12 (a)). Again, the question to be asked is: what will be the reading of a voltmeter connected in parallel with such an autonomous junction? The answer is suggested by an equivalent circuit of such a junction (Fig. 7.12 (b)) which has been constructed by adding a source of thermo-emf $E_T = \alpha(T_1 - T_2)$ to the equivalent circuit of the Josephson junction shown in Fig. 4.5. Here α is the absolute thermopower of the material of the normal layer and T_1 and T_2 are the temperatures of the superconducting electrodes where they make contact with the normal layer. It is clear from Fig. 7.12 (b) that the thermo-emf E_T included in the closed circuit produces a current which passes through the normal layer in the form of a normal current I_n, and through the true Josephson part of the junction in the form of a supercurrent $I_s = I_c \sin\varphi$. Here φ is the phase difference between the superconducting electron wavefunctions in the two S electrodes and I_c the critical current of the junction.

The voltage V_{AB} between the points A and B in Fig. 7.12 (b) must satisfy the equation

$$E_T - I_n R_n = V_{AB} , \tag{7.28}$$

where R_n is the resistance of the normal layer. On the other hand, Kirchhoff's first law must hold for both the point A and the point B:

$$I_n = I_s . \tag{7.29}$$

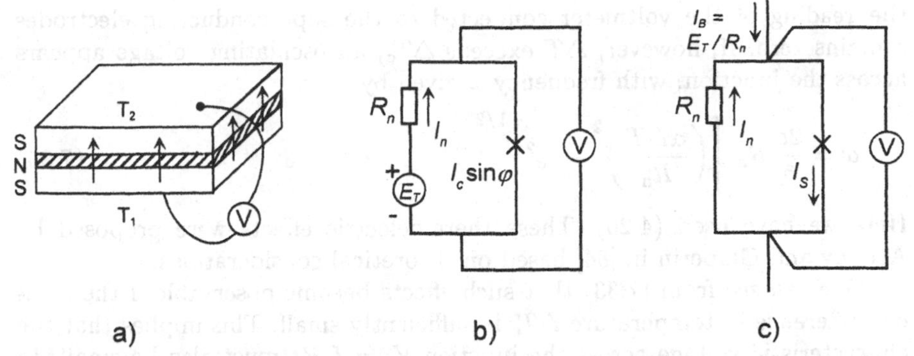

Fig. 7.12. **(a)** SNS Josephson junction with heat flow. The temperature drop across the normal layer is $(T_1 - T_2)$; **(b)** equivalent circuit for such a junction: E_T is the thermo-emf due to the heat flow across the normal layer; **(c)** the circuit corresponding to (7.32)

Finally, the part of the circuit carrying the Josephson current ought to obey the Josephson relations:

$$I_s = I_c \sin\varphi \,, \tag{7.30}$$

$$2eV_{AB} = \hbar \frac{d\varphi}{dt} \,. \tag{7.31}$$

Together, equations (7.28–31) give a complete definition of the problem. They can be easily reduced to

$$\frac{\hbar}{2eR_n}\frac{d\varphi}{dt} + I_c \sin\varphi = \frac{E_T}{R_n} \,. \tag{7.32}$$

But the last equation is equivalent to the equation of the Josephson junction in the resistively shunted model (which is already known to us from Chap. 4). The current applied to the Josephson junction from an external source (in our case E_T/R_n) is divided into two branches: a Josephson current $I_c \sin\varphi$ and a current through the normal layer $(\hbar/2eR_n)\, d\varphi/dt$; see Fig. 7.12 (c). The behavior of such a system with identical temperatures of the S electrodes is well known (see Sect. 4.3). If the external current is less than I_c, it flows through the superconducting (Josephson) part of the junction only and the voltage between the points A and B is zero. Hence the reading of the voltmeter will be zero, too. If, however, the current exceeds I_c, an oscillating voltage of frequency $\omega = 2e\overline{V}_{AB}/\hbar$ appears across the junction, where $\overline{V}_{AB}$ is the dc component of the oscillating voltage registered by the voltmeter.

Now we are in a position to give a complete description of the behavior of a Josephon junction placed in a temperature gradient. As long as the temperature drop between the two surfaces of the normal layer, $\Delta T = T_1 - T_2$, is less than a certain critical value

$$\Delta T_c = \frac{V_c}{\alpha} = \frac{I_c R_n}{\alpha} \,, \tag{7.33}$$

the reading of the voltmeter connected to the superconducting electrodes remains zero. If, however, ΔT exceeds ΔT_c, an oscillating voltage appears across the junction, with frequency ω given by

$$\omega = \frac{2e}{\hbar} R_n \left[\left(\frac{\alpha \Delta T}{R_n} \right)^2 - I_c{}^2 \right]^{1/2} . \tag{7.34}$$

Here we have used (4.20). These thermoelectric effects were proposed by Aronov and Galperin in [84] based on theoretical considerations.

One can see from (7.33) that such effects become observable if the critical difference in temperature ΔT_c is sufficiently small. This implies that the characteristic voltage across the junction $V_c = I_c R_n$ must also be small. In practice, it is not difficult to fabricate an SNS junction with $V_c \sim 10^{-13}$ V. With $\alpha \sim 10^{-8}$ V K^{-1}, this corresponds to $\Delta T_c \sim 10^{-5}$ K. Such a temperature difference can be fairly easily attained in experiment. Experimental studies of the thermoelectric effects in SNS Josephson junctions [85] have provided ample verification of the above ideas.

7.11.2 Thermoelectric Effects in SNS Josephson Junctions in the Presence of a Magnetic Field

Suppose that, in addition to the heat flow, our SNS Josephson junction is placed in a magnetic field parallel to the plane of the junction (the xz plane). In this case it turns out that one can witness interesting and beautiful phenomena.

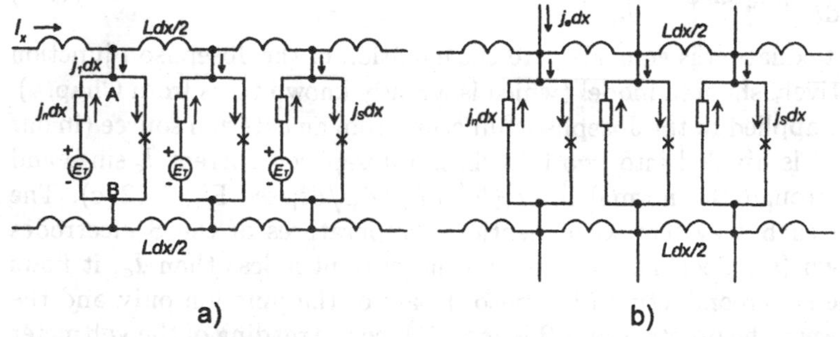

Fig. 7.13. **(a)** Equivalent circuit for a distributed SNS Josephson junction under a heat flow; **(b)** alternative representation of the same junction. The source of thermo-emf E_T is replaced by a uniform current j_e

Let us take into account the finite size of the junction, say, along the x axis. Let this size be w. The heat flow is directed along the y axis, perpendicular to the junction, and the magnetic field is in the z direction. The equivalent circuit for such a junction is shown in Fig. 7.13 (a). Here L is the total inductance of the two superconducting electrodes of the junction, per unit area, and I_x is the current in the direction of the x axis, per unit width. Now, in the same manner as we went from the equivalent circuit of Fig. 7.12 (b) to the circuit of Fig. 7.12 (c), we can go from the circuit of Fig. 7.13 (a) to the equivalent circuit of Fig. 7.13 (b). The latter is the equivalent circuit of an extended (distributed) Josephson junction of uniform temperature (that is, without heat flow) to which an external current of density $j_e = E_T/R_n$ is applied. Here R_n is the resistance per unit area of the normal layer. The behavior of such an object in an external magnetic field parallel to the y axis is well known (see Sect. 4.4). Therefore, one can foresee [86] that the critical temperature difference ΔT_c, corresponding to the appearance of a voltage across the junction, will be a nonmonotonic function of the magnetic field. In the ideal case, the dependence to be observed should have the form (see (4.46))

$$\Delta T_c = \frac{j_c R_n}{\alpha} \left| \frac{\sin(\pi\Phi/\Phi_0)}{\pi\Phi/\Phi_0} \right| , \tag{7.35}$$

where $\Phi = B_0 w d$. Here B_0 is the magnetic induction of the external field, w is the dimension of the junction along the x axis and $d = d_0 + 2\lambda_L$, with d_0 being the thickness of the normal layer.

The above prediction was verified in the following experiment [87]. The SNS Josephson junction under study was a tantalum-copper-tantalum sandwich with the thickness of the normal layer about 10^{-3} cm, the resistance of the normal layer about 10^{-9} Ω, and a critical current of about 10^{-4} A. With these parameters, the characteristic voltage across the junction, V_c, was about 10^{-13} V. The voltage V_c was measured by a SQUID voltmeter. The temperature difference between the two surfaces of the normal layer, ΔT, was produced by a heater attached to one of the superconducting electrodes of the junction and was therefore proportional to the power supplied by the heater. The critical power of the heater, P_c, was proportional to the critical temperature difference ΔT_c.

The measured critical power P_c as a function of an external magnetic field parallel to the plane of the junction is shown in Fig. 7.14 (the upper curve). The lower curve shows the dependence of the maximum zero-voltage current I_c through the junction on the external magnetic field, corresponding to a uniform temperature over the junction. The position of maximum I_c in Fig. 7.14 is shifted from $B_e = 0$ due to the effect of a background dc magnetic field. Nevertheless, when the scale of the plot was chosen in such a way that the two curves coincided at least at one point, they proved to coincide along their entire lengths. This provided convincing evidence of the validity of the above argumentation.

Let us summarize the main results of this section. Beginning from a certain critical value of the heat flow through an SNS Josephson junction, the latter starts to generate a voltage of thermoelectric nature. The critical value of the heat flow is a nonmonotonic function of the magnetic field. This phenomenon is a thermal analog of the dc Josephson effect.

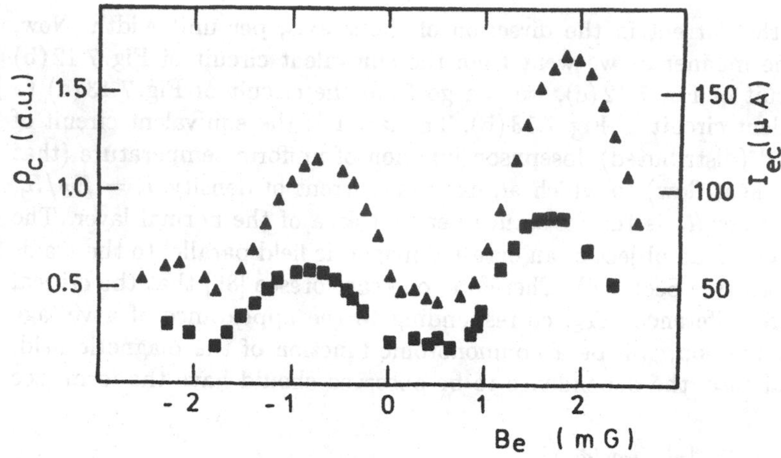

Fig. 7.14. Measurements of the magnetothermoelectric effect in an SNS Josephson junction [87]

It turned out that there also exists a thermal analog of the ac Josephson effect: a steady heat flow through an SNS Josephson junction gives rise to Josephson radiation, that is, to an oscillating voltage across the junction. This effect was observed experimentally [88] two years after the first edition of this book was published in Russian. Later on, several related thermoelectric effects, due to superconducting phase slippage, were observed in the new high-T_c materials. The most recent review on this topic is that of Huebener [81].

Problems

Problem 7.1. Find the relaxation time for the quasiparticle charge in Sn if the temperature of the material is 0.01 K below the critical temperature T_c ($T_c = 3.8$ K).

Problem 7.2. Proceeding from the requirement of electroneutrality, we have found that there exists an electric field E in the vicinity of the interface between a superconductor and a normal metal, provided there is an electric current passing through the interface. According to (7.18), div $\boldsymbol{E} \neq 0$. But this suggests that there must exist a finite volume density of the electric charge ρ, in a region of size $\sim \lambda_Q$ near the interface. The charge density is defined by Poisson's equation div $\boldsymbol{E} = 4\pi\rho$. Does this contradict the initial assumption of electroneutrality?

Solutions

Chapter 1

1.1. The surface current density j_{surf} is defined by (1.6). The field H_0 should be set equal to the critical field at $T = 4.2$ K, according to (1.1). As a result we obtain $j_{surf} = 422$ A cm^{-1}.

1.2. The difference in free energy is defined by (1.12). The value of H_{cm} at 4.2 K can be calculated using (1.1). Finally, using data from Table 1.1, we have $F_n - F_{s0} = 1.1 \times 10^4$ erg cm^{-3}.

1.3. The latent heat given up by the specimen is $Q = T(S_n - S_s)$. Using (1.1) and (1.18), we obtain $Q = 2.30 \times 10^4$ erg.
Note. Formula (1.18) expresses the difference in entropy through the quantity $(\partial H_{cm}/\partial T)_R$. Under the conditions of this particular problem, when the fixed quantity is the external field at infinity, the superconductor does not do any work on external bodies when its temperature changes. Indeed, the only kind of work on external bodies possible for such a superconductor is radiation of electromagnetic waves, which may occur as a result of a change in its internal state. But this is only possible if the integral of the Poynting vector $(c/4\pi)[\boldsymbol{E} \times \boldsymbol{H}]$ over the surface of the superconductor is nonzero. An electric field $\boldsymbol{E}$ can only appear as a result of variations in the magnetic induction $\boldsymbol{B}$, but $\boldsymbol{B} = 0$. This implies that a change in temperature of a superconductor placed in a magnetic field cannot be accompanied by electromagnetic radiation and $(\partial H_{cm}/\partial T)_R = (\partial H_{cm}/\partial T)_H$. That is, we can indeed use (1.18).

1.4. Searching for a minimum of the entropy difference (1.18), we find that the temperature in question is $T_c/\sqrt{3} = 4.16$ K.

1.5. 108 Oe.

1.6. 2.8×10^4 erg cm^{-3}.

Chapter 2

2.1. $H = 1.84 \times 10^{-2}$ Oe.

2.2. $A = 75$ G cm; $\nabla\theta = 2.36 \times 10^9$ rad cm^{-1}.

2.3. Since $d \ll \lambda$, the current distribution over the film is uniform. Therefore, integrating (2.32) over a closed circular contour of radius R yields

$$\Phi = \Phi_0 n - \frac{2\pi mcR}{n_s e^2} j_s \,. \tag{S.1}$$

On the other hand, the magnetic field in the interior of the cylinder and the current at its surface are related to each other by

$$H = \frac{4\pi}{c} j_s d \,,$$

so that the magnetic flux through the cylinder is $\Phi = (4\pi^2/c)\, j_s dR^2$. Using this to express j_s and substituting it into (S.1) finally yields

$$\Phi = \Phi_0 n \left(1 + \frac{2\lambda^2}{Rd}\right)^{-1} .$$

Note. At $Rd \gg 2\lambda^2$, quantization of the magnetic flux through a thin-walled cylinder ($d \ll \lambda$) is totally analogous to that through a bulk cylinder. In all other cases, the 'flux quantum' is less than Φ_0.

2.4. See Fig. S.1.

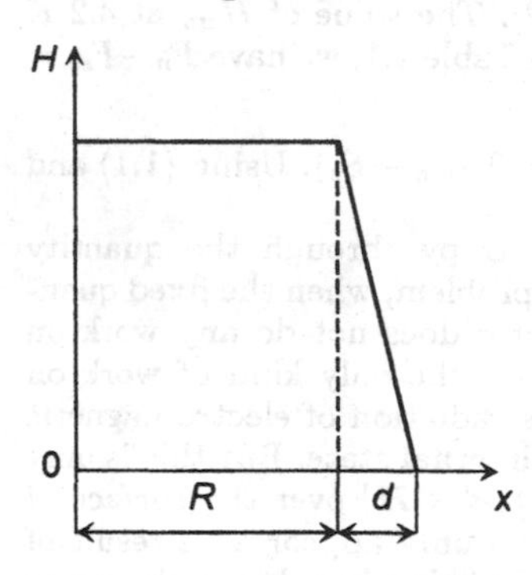

Fig. S.1. Distribution of the magnetic field trapped by a thin-walled superconducting cylinder

Chapter 3

3.1. If $\lambda(0) = 390$ Å, we have $\lambda(4.2\,\text{K}) = 415$ Å. The magnetic field at a distance 300 Å from the surface will then be $H = H_0\, e^{-x/\lambda} = 145.6$ Oe and $w = H^2/8\pi = 844\ \text{erg cm}^{-3}$.

3.2. $\lambda(7.10\,\text{K})/\lambda(4.2\,\text{K}) = 4.49$; $n_s(7.10\,\text{K}) = 8.06 \times 10^{20}\ \text{cm}^{-3}$.

3.3. Since $\lambda \ll \xi$, we can use (3.56). Finding H_{cm} from (3.38) and substituting it into (3.56), we obtain $\sigma_{\text{ns}} = 1.24 \times 10^{-2}\ \text{erg cm}^{-2}$.

3.4. Since $\lambda(T = 0.9\,T_c) = 870$ Å (see Problem 3.3) we have, according to (3.61), $H(x = 0) = 8.55$ Oe. Here we have assumed that the magnetic field is sufficiently small ($H_0 \ll H_{\text{cm}}(T = 0.9\,T_c) \approx 60$ Oe) so that the order parameter ($\psi = 1$) remains unaffected.

The density of the diamagnetic moment is $M(x) = (1/4\pi)[H(x) - H_0]$. Taking, according to (3.61), $H(x) = H_0 \cosh(x/\lambda)/\sinh(x/\lambda)$ and integrating the diamagnetic moment over the thickness of the film, we find M_0:

$$M_0 = \frac{H_0 d}{4\pi}\left(\frac{2\lambda}{d}\tanh\frac{d}{2\lambda} - 1\right). \qquad \text{(S.2)}$$

It follows that $M_0 = -7.74 \times 10^{-7}$ G cm.

3.5. It follows from (S.2) that the average density of the diamagnetic moment is $\overline{M} = (1/4\pi)\, H_0\, [(2\lambda/d)\tanh(d/2\lambda) - 1]$. The work done by the source of magnetic field equals

$$W(H_0) = -\int_0^{H_0} \overline{M}\, \mathrm{d}H_0 \,.$$

The transition to the normal state occurs when this work becomes equal to $F_n - F_{s0} = H_{cm}^2/(8\pi)$. Therefore

$$H_{c\parallel} = H_{cm}\left(1 - \frac{2\lambda}{d}\tanh\frac{d}{2\lambda}\right)^{-1/2}.$$

Recalling that $\lambda \ll d$, we have $H_{c\parallel} = H_{cm}(1 + \lambda/d) = 1.087\, H_{cm}$.

3.6. $H_{I_c} = 12.5$ Oe; $H_{c\parallel} = 1015$ Oe.

3.7. $j_c = 2.30 \times 10^8$ A cm^{-2}.

3.8. The critical current density is defined by (3.67) and the critical velocity by (3.69). Using (2.7) and (3.37), we have

$$v_c = \frac{\hbar\kappa}{3\sqrt{3}\, m\lambda} = 2.81 \times 10^3 \text{ cm s}^{-1}\,.$$

3.9. According to (2.22), $j_M = cH_{cm}/4\pi\lambda$ and $j_M/j_c = 3\sqrt{3}/2\sqrt{2} = 1.837$. This means that, if the field at the surface is equal to H_{cm}, the density of the Meissner current exceeds the pair-breaking current by 1.84 times. This happens because (2.22) does not take into account the decrease of the superconducting electron density n_s caused by the magnetic field H_{cm}.

Chapter 4

4.1. Since the junctions are connected in parallel, the phase differences φ_1 and φ_2 across them are identical and the currents through the junctions are distributed in proportion to their critical currents: $I_1 = 0.417$ mA, $I_2 = 0.583$ mA.

4.2. $V_{max} - V_{min} = 2I_c R$.

4.3. $\overline{V} = 1.33$ mV; $\nu = 641$ GHz.

4.4. We seek the phase difference across the junction in the form $\varphi = \varphi_0 + \varphi_1$, where φ_0 is the phase difference due to the dc current I_0. The voltage across the junction is $V = (\Phi_0/2\pi)\, \mathrm{d}\varphi/\mathrm{d}t = (\Phi_0/2\pi)\, \mathrm{d}\varphi_1/\mathrm{d}t$, where $\mathrm{d}\varphi_1/\mathrm{d}t = (\mathrm{d}I/\mathrm{d}t)\,(\mathrm{d}I/\mathrm{d}\varphi_1)^{-1}$. Substituting this expression into the expression for V, we obtain

$$V = \frac{\Phi_0}{I_c \cos\varphi_0}\, \nu I_1 \sin(2\pi\nu t + \pi/2).$$

With the parameters given in the problem, the amplitude of the ac voltage will be 0.58 nV.

Note. It follows from the above solution that there is a positive phase angle of $\pi/2$ between the voltage V and the current through the junction. This implies that the junction behaves as an inductance $L = \Phi_0/(2\pi I_c \cos\varphi_0)$.

4.5. $H_{c1} = 0.290$ Oe; $H(0) = 0.455$ Oe.

4.6. The first maximum will appear at $H_1 = 1.85$ Oe and the second at $H_2 = 3.18$ Oe.

4.7. The dependence of Φ on Φ_e will not show hysteresis if the critical current of the junction is less than $\Phi_0/(2\pi L) = 0.165 \times 10^{-6}$ A.

4.8. The current in the ring is determined by the phase difference φ across the junction: $\varphi = 2\pi L I_R/\Phi_0$. The current through the junction is $I_J = I_c \sin\varphi$. The total current is $I = 1.34 \times 10^{-6}$ A.

4.9. $\varphi_{AB} = 1.60$ rad, see Fig. S.2.

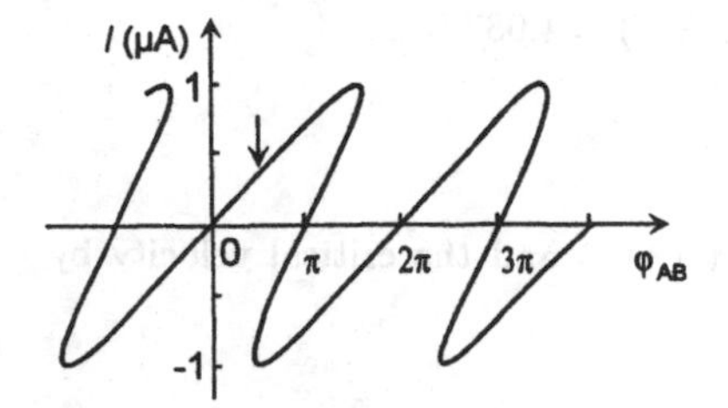

Fig. S.2. Answer to Problem 4.9. Current as a function of the phase difference between points A and B. The arrow marks the phase difference sought, $\varphi_{AB} = 1.60$ rad

Chapter 5

5.1. The magnetic field generated by one of the vortices at a distance $r \ll \lambda$ from its center is, according to (5.6) and (5.7),

$$H = \frac{\Phi_0}{2\pi\lambda^2} \ln\frac{\lambda}{r} .$$

The density of the current generated by the vortex at the same point is

$$j = \frac{c}{4\pi} \frac{\Phi_0}{2\pi\lambda^2 r} .$$

Finally, using (5.23), we find that the vortex will experience a repulsive force

$$f = \frac{\Phi_0{}^2}{4\pi^2\lambda^2} \frac{x}{x^2 + (d/2)^2}$$

directed away from the line ab, along the dashed line in Fig. 5.22.

5.2. $\lambda = \dfrac{1}{H_{cm}} \sqrt{\dfrac{\Phi_0 H_{c2}}{4\pi}} = 1.48 \times 10^{-5}$ cm.

5.3. $H_{c1} = 244$ Oe.

5.4. Taking into account that $\xi \ll \lambda$ and using (5.6) and (3.38), we find that the field at a distance $r \sim \xi$ is

$$H(r) = \frac{\sqrt{2}H_{\rm cm}}{\kappa} K_0\left(\frac{r}{\lambda}\right) .$$

The current density will then be $j = (c/4\pi)\, \mathrm{d}H/\mathrm{d}r$. This gives

$$|j| = \frac{\sqrt{2}}{4\pi} \frac{cH_{\rm cm}}{\lambda} .$$

Note. It follows from the above that, by the order of magnitude, the current density at a distance ξ from the center of the vortex is equal to the pair-breaking current (compare with (3.67)).

5.5. Knowing H_{c2}, we find $\xi = 4.69 \times 10^{-7}$ cm and knowing κ, we find $\lambda = 4.50 \times 10^{-5}$ cm. Further, using (5.12), we find $\varepsilon = 6.12 \times 10^{-7}$ erg cm^{-1}. The vortex core is normal, which implies that its energy density exceeds that of the surrounding medium by $H_{\rm cm}^2/8\pi$. Therefore, the energy of the core equals $H_{\rm cm}^2\xi^2/8$, if its radius is assumed equal to ξ. With the parameter values of our problem, the energy is 3.33×10^{-8} erg cm^{-1}, that is, many times less than the electromagnetic energy of the vortex, ε.

5.6. $M = -0.445$ G.

5.7. It is evident from Fig. 5.23 that H_{c1} is given by the solution of the following equation: $H_{c1} = k\,(H_{c2} - H_{c1})$, where $k^{-1} = 1.16\,(2\kappa^2 - 1)$. Then $H_{c1} = 193$ Oe. Further, an evaluation of $H_{\rm cm}$ can be obtained from the area enclosed by the magnetization curve, as follows from (5.35). This gives $H_{\rm cm} = [k/(1+k)]^{1/2} H_{c2} = 879$ Oe. The exact value of $H_{\rm cm}$ can be found using (5.25): $H_{\rm cm} = 942$ Oe.

5.8. $H_{c3} = 51\,900$ Oe.

5.9. $H_{c3} = 68\,700$ Oe.

5.10. $f = \dfrac{(\Phi_0/4\pi\lambda)^2}{l} = 6.03 \times 10^{-2}$ dyn cm^{-2}.

5.11. $f_x = -0.208$ dyn cm^{-1}; $f_y = -6.20 \times 10^{-2}$ dyn cm^{-1}.

5.12. $f = \dfrac{\Phi_0 H_0}{4\pi\lambda}\, \mathrm{e}^{-l/\lambda} = 5.52 \times 10^{-2}$ dyn cm^{-1}.

5.13. $H_{\rm p} = \dfrac{\Phi_0 \kappa}{2\sqrt{2}\pi\lambda^2} = 1.40 \times 10^3$ Oe.

5.14. Using (5.17) and (5.26), we find $\kappa = 31$. From (5.17) we find $\lambda = 2.1 \times 10^{-5}$ cm. Finally, approximating the magnetization curve $|M|$ by two triangles (see Fig. 5.23) and using (5.57), we have

$$I_{\rm c} = \frac{2c\sqrt{\Phi_0}\, l}{\lambda\sqrt{B}} \frac{H_{c1}}{4\pi} \frac{H_{c2} - B}{H_{c2} - H_{c1}} \approx 13 \text{ A} .$$

5.15. It follows from (5.67) and (5.68) that $\rho_{\rm f} = \rho_{\rm n} B/H_{c2}(0)$. Taking approximately $H_{c2}(T) = H_{c2}(0)\,(1 - T^2/T_{\rm c}{}^2)$ and $B = H_0$, we find $\rho_{\rm f} = 3 \times 10^{-6}\ \Omega$ cm.

5.16. $v_{\rm L} = \dfrac{U}{lB} = 3.75 \times 10^{-5}$ m s^{-1}.

Chapter 6

6.1. $\rho/N(0) = 7.12;\ 1.34;\ 1.15.$

6.2. Using (6.37), we obtain $g = 0.245$.

6.3. The critical temperature will decrease by 0.01 K.

6.4. $T_c = \dfrac{2\Delta_0}{3.52k_B} = 3.49\ \mathrm{K}$.

6.5. $K_c = \dfrac{2\Delta_0}{\hbar v_F} = 2.61 \times 10^4\ \mathrm{cm}^{-1}$, $k_F = \dfrac{mv_F}{\hbar} = 5.6 \times 10^7\ \mathrm{cm}^{-1}$.

Chapter 7

7.1. Using the expression for the gap (at $T \approx T_c$): $\Delta(T) = 1.74\,\Delta(0)\,(1 - T/T_c)^{1/2}$, we obtain for Sn: $\tau_Q \approx 2.4 \times 10^{-9}$ s.

7.2. Let us show that the volume charge density ρ, which will indeed appear in the transitional region of width $\sim \lambda_Q$ at the interface, will be many orders of magnitude less than the volume charge density of the quasiparticles, eQ. Poisson's equation can be written approximately as $E/\lambda_Q \sim 4\pi\rho$, or $\rho \sim E/4\pi\lambda_Q$. On the other hand, we have from (7.12):

$$Q \sim \frac{j_n \tau_Q}{e\lambda_Q} = \frac{\sigma\tau_Q E}{e\lambda_Q}\,,$$

which leads to

$$\frac{\rho}{eQ} \sim \frac{1}{4\pi\tau_Q\sigma} = \left(\frac{8\pi\tau_Q e^2\,N(0)\,l v_F}{3}\right)^{-1}.$$

With the values $\tau_Q \approx 10^{-10}$ s, $N(0) \approx 10^{33}\ \mathrm{erg}^{-1}\,\mathrm{cm}^{-1}$, $l \approx 10^{-8}$ cm, and $v_F \approx 10^8\ \mathrm{cm\,s}^{-1}$, we obtain $\rho/eQ \sim 10^{-5}$.

References

1. A.C. Rose-Innes, E.H. Rhoderick: *Introduction to Superconductivity* (Pergamon Press, Oxford 1978)
2. E.A. Lynton: *Superconductivity* (Methuen, London 1962)
3. M. Tinkham: *Introduction to Superconductivity* (MacGraw-Hill, New York 1996)
4. P.G. de Gennes: *Superconductivity of Metals and Alloys* (Benjamin, New York 1966)
5. D.R. Tilley, J. Tilley: *Superfluidity and Superconductivity* (Hilger, Bristol 1986)
6. H. Kamerlingh Onnes: Leiden Comm. **122b**, 124 (1911)
7. B.W. Roberts: J. Phys. Chem. Ref. Data **5**, 581 (1976)
8. B.S. Deaver, Jr., W.M. Fairbank: Phys. Rev. Lett. **7**, 43 (1961)
9. R. Doll, M. Näbauer: Phys. Rev. Lett. **7**, 51 (1961)
10. B.D. Josephson: Phys. Lett. **1**, 251 (1962)
11. I.K. Yanson, V.M. Svistunov, I.M. Dmitrenko: Zh. Eksp. Teor. Fiz. **48**, 976 (1965) [English transl.: Sov. Phys. JETP **21**, 650 (1965)]
12. W. Meissner, R. Ochsenfeld: Naturwiss. **21**, 787 (1933)
13. L.D. Landau, E.M. Lifshitz: *Electrodynamics of Continuous Media* (Nauka, Moscow 1972) [English transl.: 2nd edn., Pergamon Press, Oxford 1968]
14. A.G. Meshkovskii, A.I. Shalnikov: Zh. Eksp. Teor. Fiz. **17**, 851 (1947)
15. A.G. Meshkovskii, A.I. Shalnikov: Zh. Eksp. Teor. Fiz. **19**, 1 (1949)
16. A. Bodmer, U. Essmann, H. Traüble: Phys. Status Solidi (a) **13**, 471 (1972)
17. F. London, H. London: Proc. Roy. Soc. **A149**, 71 (1935)
18. A.B. Pippard: Proc. Roy. Soc. **A216**, 547 (1953)
19. V.L. Newhouse, J.W. Bremer, H.H. Edwards: Proc. IRE **48**, 1395 (1960)
20. K.K. Likharev, B.T. Ulrich: *Systems with Josephson Junctions. The Basics of the Theory* (Moscow State Univ. Press, Moscow 1978)
21. T. van Duzer, C.W. Turner: *Principles of Superconductive Devices and Circuits* (Elsevier, New York 1981)
22. J.I. Gittlemann, B. Rosenblum: Proc. IEEE **52**, 1138 (1964)
23. V.L. Ginzburg, L.D. Landau: Zh. Eksp. Teor. Fiz. **20**, 1064 (1950)
24. L.D. Landau, E.M. Lifshitz: *Statistical Physics*, 3rd edn., part 1 (Nauka, Moscow 1976) [English transl.: Pergamon Press, Oxford 1980]
25. K.K. Likharev, L.A. Yakobson: Zh. Teor. Fiz. **45**, 1503 (1975)
26. G. Deutscher, P.G. de Gennes: In *Superconductivity*, ed. by R.D. Parks, Vol. 2, p.1005 (Marcel Dekker Inc., New York 1969)
27. Z.G. Ivanov, M.Yu. Kupriyanov, K.K. Likharev, O.V. Snigirev: J. de Physique **C-6**, 556 (1978)
28. J.J. Hauser, H.C. Theuerer: Phys. Lett. **14**, 270 (1965)
29. N.V. Zavaritzkii: Doklady Akad. Nauk SSSR **78**, 665 (1951)
30. S. Shapiro: Phys. Rev. Lett. **11**, 80 (1963)
31. K.K. Likharev: Rev. Mod. Phys. **51**, 101 (1979)

32. L. Solymar: *Superconductive Tunnelling and Applications* (Chapman and Hall, London 1972)
33. R. Gross: In *Interfaces in Superconducting Systems*, ed. by S.L. Shinde and D. Rudman, pp. 176–209 (Springer, New York 1994)
34. R.P. Feynman, R.B. Leighton, M. Sands: *The Feynman Lectures on Physics. 3: Quantum Mechanics* (Addison-Wesley, Reading, MA 1965)
35. L.G. Aslamazov, A.I. Larkin: Pis'ma Zh. Eksp. Teor. Fiz. **48**, 976 (1965)
36. R. Ferrel, R. Prange: Phys. Rev. Lett. **10**, 479 (1963)
37. D.W. McLaughlin, A.C. Scott: Phys. Rev. A **18**, 1652 (1978)
38. N.F. Pedersen: In *Solitons*, ed. by S.E. Trullinger, V.E. Zakharov, and V.L. Pokrovsky, p. 469 (Elsevier, Amsterdam 1986)
39. R.D. Parmentier: In *The New Superconducting Electronics*, ed. by H. Weinstock, R.W. Ralston, p. 221 (Kluwer, Dordrecht 1993)
40. D.N. Langenberg, D.J. Scalapino, B.N. Taylor: Proc. IEEE **54**, 560 (1966)
41. J. Clarke: In *Superconductor Applications: SQUIDs and Machines*, ed. by B.B. Schwartz and S. Foner (Plenum Press, New York 1977)
42. J. Clarke: In *Superconducting Electronics*, ed. by H. Weinstock and M. Nisenoff (NATO ASI Series) **F59**, 87–148 (1989); **E251**, 165 (1993)
43. H.J. Hartfuss, K.H. Gundlach, V.V. Schmidt: J. Appl. Phys. **52**, 5411 (1981)
44. J. Niemeyer: In *Superconducting Quantum Electronics*, ed. by V. Kose, pp. 228–254 (Springer, Berlin 1989)
45. See articles in: *Macroscopic Quantum Phenomena and Coherence in Superconducting Networks*, ed. by C. Giovannella and M. Tinkham (World Scientific, Singapore 1995)
46. K.K. Likharev, V.K. Semenov: IEEE Trans. Appl. Supercond. **1**, 3 (1991)
47. A.A. Abrikosov: Zh. Eksp. Teor. Fiz. **32**, 1442 (1957) [English transl.: Sov. Phys. JETP **5**, 1174 (1957)]
48. U. Essmann, H. Träuble: Phys. Lett. A **24**, 526 (1967)
49. D. Saint-James, P.G. de Gennes: Phys. Lett. **7**, 306 (1963)
50. D. Saint-James, G. Sarma, E.J. Thomas: *Type II Superconductivity* (Pergamon Press, Oxford 1969)
51. C.P. Bean, J.D. Livingston: Phys. Rev. Lett. **12**, 14 (1964)
52. R.W. de Blois, W. de Sorbo: Phys. Rev. Lett. **12**, 499 (1964)
53. W.E. Lawrence, S. Doniach: In *Proc. of the 12th Int. Conf. on Low Temperature Physics*, ed. by E. Kanda, p.361 (Tokyo 1970)
54. E.H. Brandt: J. Low Temp. Phys. **709** (1977); Phys. Stat. Sol. (b) **77**, 551 (1976)
55. A.M. Campbell, J.E. Evetts, D. Dew-Hughes: Phil. Mag. **18**, 313 (1968)
56. G.S. Mkrtchyan, V.V. Schmidt: Zh. Eksp. Teor. Fiz. **61**, 367 (1971) [English transl.: Sov. Phys. JETP **34**, 195 (1972)]
57. M. Kraus: Thesis (University of Erlangen–Nürnberg 1995)
58. M. Leghissa, L.A. Gurevitch, M. Kraus, G. Saemann-Ischenko, L.Ya. Vinnikov: Phys. Rev. B **48**, 1341 (1993)
59. K.E. Osborn, A.C. Rose-Innes: Phil. Mag. **27**, 683 (1973)
60. I.O. Kulik, I.K. Yanson: *Josephson Effect in Superconducting Tunneling Structures* (Nauka, Moscow 1970)
61. P. Martinoli: Phys. Rev. B **17**, 1175 (1978)
62. J. Bardeen, L. Cooper, J. Schrieffer: In *Theory of Superconductivity*, ed. by N.N. Bogolyubov, p. 103 (Inostrannaya Literatura, Moscow 1960)
63. I. Giaever: Phys. Rev. Lett. **5**, 147 (1960)
64. I. Giaever, K. Megerle: Phys. Rev. **122**, 1101 (1961)
65. F. London: Proc. Roy. Soc. **A152**, 24 (1935); Phys. Rev. **74**, 562 (1948)

66. L.P. Gorkov: Zh. Eksp. Teor. Fiz. **36**, 1918 (1959) [English transl.: Sov. Phys. JETP **9**, 1364 (1959)]
67. C.J. Pethick, H. Smith: Annals of Physics **119**, 133 (1979)
68. E.M. Lifshtz, L.P. Pitaevskii: *Physical Kinetics* (Pergamon Press, Oxford 1981)
69. A. Schmid, G. Schön: J. Low Temp. Phys. **20**, 207 (1975)
70. A.F. Andreev: Zh. Eksp. Teor. Fiz. **46**, 1823 (1964) [English transl.: Sov. Phys. JETP **19**, 1228 (1964)]
71. M.L. Yu, J.E. Mercereau: Phys. Rev. B **12**, 4909 (1975)
72. V.V. Ryazanov, L.A. Ermolaeva, V.V. Schmidt: J. Low Temp. Phys. **45**, 507 (1981)
73. S.N. Artemenko, A.F. Volkov: Uspekhi Fiz. Nauk **128**, 3 (1979) [English transl.: Sov. Phys. Uspekhi **22**, 295 (1979)]
74. T.Y. Hsiang, J. Clarke: Phys. Rev. B **21**, 945 (1980)
75. J. Meyer, G. v. Minnegerode: Phys. Lett. A **38**, 529 (1972)
76. W.J. Skocpol, M.R. Beasley, M. Tinkham: J. Low Temp. Phys. **16**, 145 (1974)
77. A.F.G. Wyatt, V.M. Dmitriev, W.S. Moore, F.W. Sheard: Phys. Rev. Lett. **16**, 1166 (1966)
78. G.M. Eliashberg: J. Low Temp. Phys. **10**, 449 (1973)
79. V.L. Ginzburg, G.F. Zharkov: Uspekhi Fiz. Nauk **125**, 19 (1978) [English transl.: Sov. Phys. Uspekhi **21**, 381 (1978)]
80. D.J. van Harlingen: Physica B **109–110**, 1710 (1982)
81. R.P. Huebener: Supercond. Sci. Technol. **8**, 189 (1995)
82. N.V. Zavaritskii: Pis'ma Zh. Eksp. Teor. Fiz. **19**, 205 (1974) [English transl.: JETP Lett. **19**, 126 (1974)]
83. D.J. van Harlingen, D.F. Heidel, J.C. Garland: Phys. Rev. B **21**, 1842 (1980)
84. A.G. Aronov, Yu.M. Gal'perin: Pis'ma Zh. Eksp. Teor. Fiz. **19**, 281 (1974) [English transl.: JETP Lett. **19**, 165 (1974)]
85. M.V. Kartsovnik, V.V. Ryazanov, V.V. Schmidt: Pis'ma Zh. Eksp. Teor. Fiz. **33**, 373 (1981) [English transl.: JETP Lett. **33**, 356 (1981)]
86. V.V. Schmidt: Pis'ma Zh. Eksp. Teor. Fiz. **33**, 104 (1981) [English transl.: JETP Lett. **33**, 98 (1981)]
87. V.V. Ryazanov, V.V. Schmidt: Solid State Commun. **40**, 1055 (1981)
88. G.I. Panaitov, V.V. Ryazanov, A.V. Ustinov, V.V. Schmidt: Phys. Lett. A **100**, 301 (1984)

Index

Springer and the environment

At Springer we firmly believe that an international science publisher has a special obligation to the environment, and our corporate policies consistently reflect this conviction.
We also expect our business partners – paper mills, printers, packaging manufacturers, etc. – to commit themselves to using materials and production processes that do not harm the environment. The paper in this book is made from low- or no-chlorine pulp and is acid free, in conformance with international standards for paper permanency.